T0329827

Innovation Strategies for a Global Economy

To
Charlotte, David and Elizabeth

Innovation Strategies for a Global Economy

Development, Implementation, Measurement and Management

Fred Gault

Professorial Fellow, UNU MERIT, The Netherlands and Professor Extraordinaire and Member of the TUT Institute for Economic Research on Innovation (IERI), Tshwane University of Technology (TUT), South Africa

Edward Elgar
Cheltenham, UK • Northampton, MA, USA

International Development Research Centre
Ottawa • Cairo • Dakar • Montevideo • Nairobi • New Delhi • Singapore

Copublished by
Edward Elgar Publishing Limited
The Lypiatts
15 Lansdown Road
Cheltenham
Glos GL50 2JA
UK

Edward Elgar Publishing, Inc.
William Pratt House
9 Dewey Court
Northampton
Massachusetts 01060
USA

International Development Research Centre
PO Box 8500
Ottawa, ON, K1G 3H9
Canada
www.idrc.ca / info@idrc.ca
ISBN 978-1-55250-484-0 (IDRC e-book)

A catalogue record for this book
is available from the British Library

Library of Congress Control Number: 2009942846

Mixed Sources
Product group from well-managed
forests and other controlled sources
www.fsc.org Cert no. SA-COC-1565
© 1996 Forest Stewardship Council

ISBN 978 1 84980 036 5 (cased)

Printed and bound by MPG Books Group, UK

Contents

Tables and boxes

TABLES

BOXES

Preface

When I first came to the Organisation for Economic Co-operation and Development (OECD) in Paris as a delegate to the Working Party of National Experts on Science and Technology Indicators (NESTI) I was struck by the breadth of the agenda and the mix of policy analysts and statistical experts at the table able to interact effectively and to speak a common technical language then rooted in the Frascati Manual.

When the discussions began on how to measure the activity of innovation, at a time when the subject of innovation studies was emerging I would have welcomed a guide to the measurement issues, the policy issues and how they all fitted together. That was the first glimmer of what became this book.

As the years passed, I heard the need for a practitioners' guide to the subject echoed by new members not just of NESTI but of other committees in the domain of science, technology and innovation. The policy committees have working parties attached to them that deal with more focused matters of policy or statistical measurement. In that sense NESTI is subordinate to the Committee for Scientific and Technological Policy (CSTP). Viewed differently, NESTI is the current incarnation of an expert committee that met before there was an OECD in 1957 and which gave rise to the first edition of the Frascati Manual, drafted by Christopher Freeman, which governed the collection and interpretation of research and development (R&D) data, as the sixth edition does in 2009.

In 2008 I was invited to become a Visiting Fellow at Canada's International Development Research Centre (IDRC), and one of the expectations was that I would use the time to write a book on innovation measurement and policy that would contribute to the discourse in the innovation community and would also serve as a teaching tool for courses the IDRC planned for delivery in developing countries. This was a great opportunity and made more so by an invitation from the OECD to become a member of the management team working on the OECD Innovation Strategy to be delivered in June 2010. This fitted well with IDRC and with my work with the New Partnership for Africa's Development (NEPAD) Office of Science and Technology. It also gave rise to a problem of using privileged information in the preparation of this book. The problem has been resolved by using sources that are in the public domain and completing the book before the Innovation Strategy policy principles and policy advice have been decided. Any suggestions for future action in the book are mine, and not those of the IDRC, or the OECD, and any benefits from the publication of the book have been assigned to the IDRC. It will be used as a teaching tool.

In capitals of OECD countries, there are departments of education and research, of innovation, of science and technology and others with titles including various combinations of those words. In the same capitals are statistical offices responsible for producing the statistics for the System of National Accounts (SNA), the best-known SNA indicator, of course, is the gross domestic product (GDP). The GDP has a much longer history than indicators of R&D, such as gross domestic expenditure on R&D (GERD) which, in turn, has been around longer than the propensity to innovate by firms. Some statistical offices venture into the realm of science, technology and innovation (STI) statistics, which are done on the margin of the SNA. They then publish the GERD/GDP ratio which the OECD tabulates for member countries and some others. This is a league table which encourages political leaders to set targets for their country in terms of R&D performed.

Not all official statistics come from statistical offices. Some are gathered by policy departments and some from industry associations or research institutes. However this is done, there are those who produce estimates for indicators and those who use the indicators for policy purposes. The European Union target of GERD/GDP ratio of 3 per cent (with 2 per cent from the business sector) is an example, as is the new US target of more than 3 per cent set by President Obama on 28 April 2009.

While there are policy-makers and statisticians, they each have a limited understanding of what the other does. Addressing this is one of the objectives of this book. If the understanding increases, the likelihood of misinterpretation of statistics declines and their use in the policy process becomes more effective. The statistician will begin to understand the phrase, 'The Minister wants it *now*', and the policy person will understand the reason for the response of: 'We cannot give you the R&D expenditures of the top five firms in the country.'

The book is about innovation, its measurement, the use of indicators in policy and the policy learning that results. There are some recurring themes. One is that indicators of innovation are not much used in policy. That is not an original observation, but it is a disturbing one. It may be that innovation indicators are too new and that politicians and senior bureaucrats are not yet ready to use them. It may also be that there is no one indicator that really describes the innovation system of a country, and that makes it difficult to produce a sound bite. The closest would be the propensity of a firm to innovate in a particular sector or industry, but what does that mean? The qualified answer to that can be found in the book.

Another theme, much commented upon by my colleague Anthony Arundel at the United Nations University Maastricht Economic and Social Research and Training Centre (UNU-MERIT), is that more firms are innovating than are doing R&D. This is one of the more robust results to come out of many years of surveying. It even appears in a footnote of the first edition of the Oslo Manual published in 1992. But where are the policies that help firms that have to solve problems to survive, but which do no R&D and cannot claim R&D tax credits or other R&D-related support? The suggestion in the text is that if these firms were supported, and growth was one of the objectives of the support package, they might in fact grow. Larger firms have a higher propensity to do R&D. This is an indirect approach to getting to the 3 per cent targets. It also makes the point that size of firm is a key variable for analysis.

Then there are the users of products who feel obliged to change them to make them suit their needs better; or if they have a need which is not being met by the market, they develop their own products to meet their needs. These can be consumers but they can also be firms that need to improve their production process to put products on the market. These firms are not selling production processes, but they have to make them work to survive.

This introduces another historical theme in the book. In 1987 a senior member of the Canadian statistical office, Statistics Canada visited the US Census Bureau where Gaylord Worden presented a survey questionnaire on the use and planned use of advanced manufacturing technologies. This was an inspired questionnaire that took some years to develop, but was easy for a plant manager to fill out. A copy of the questionnaire found its way to Ottawa and a pilot survey was done and published on Thursday, 15 October 1987. The Canadian press published the story the following day and the market crashed on 19 October, Black Monday. There was no causal relationship, but competitiveness became a policy goal in the downturn and a decision was taken to do a full survey and for me to work with Robert Tinari at the US Census Bureau to produce the first, and the last, Canada–US comparison of the use of advanced manufacturing technologies in five industries. In all five, Canada did not do as well as the US. This result provided motivation for work at the Canadian department responsible for industry policy for some years.

After the 1987 survey, Eric von Hippel published his 1988 book which identified the user, or consumer, as the source of much innovation. As a result of this, the manager of the 1989 survey, Louis Marc Ducharme, added a question on the modification of the technologies that had been adopted, by the users – the first measure of user innovation in official statistics that produced population estimates of this activity. The population of technology adopters that went on to modify the technologies was significant.

When the technology use survey was repeated in 1998 there were questions about technology modification and about the development of technologies by users that were not available on the market to solve their production problems. Both activities were significant. When a technology use survey was conducted for 2007, the same questions were there, and this time there was a pilot follow-up survey which asked a number of questions about how the users that modified or developed their technologies funded the activity and protected the intellectual property that resulted. The interesting finding was that a significant number of user innovators chose to give the knowledge away, rather than to protect it with intellectual property instruments. That raises some questions for intellectual property policy.

So far, the themes are the bringing closer together of the policy community and the measurement community, the importance of non-R&D-performing firms to value creation and economic growth, and the significant role of the user innovator in the activity of innovation. While the three themes are important, there are two overarching themes which explain why it is necessary to have a better understanding of innovation and of what governments can do to make it work better. Those themes are avoiding ecological and financial disaster.

There are two aspects of this; the first is to deal with climate change through innovation and to save the planet and the higher life forms that inhabit it, of which the reader is an example. The second is to promote innovation that results in economic growth, rather than the economic decline now being experienced as a result of innovation in financial services, and the rapid diffusion of the monetized debt products, until they lost value and the market crashed. There are other reasons for understanding innovation, but these are high on the list.

Understanding innovation is not just a matter of running surveys and feeding the results into evidence-based policy. The innovation system is global, complex, dynamic and non-linear in its response to policy intervention. This phrase is a leitmotif in the book, along with the other themes, and it poses a challenge that is left to the next generation. However, there is a suggestion about how to proceed.

Jack Marburger came to Ottawa in 2006 to the OECD Blue Sky II Forum and put the case for the development of a science of science and innovation policy. The idea is the creation of a new cross-disciplinary social science that will improve the understanding of the science of policy, in this case the science of science and innovation policy. As this book is about innovation, the focus has been on the creation of a science of innovation policy and the book lays out the components of a research agenda that moves in the direction of a new science.

Finally, the book comes at a time when the OECD is developing its Innovation Strategy, as is the European Union (EU), which is considering an Innovation Act in 2010. These events should be seen not as an end but as step towards longer-term goals, to which this book is a contribution.

Acknowledgements

The idea for this book goes back many years but the opportunity to write it was provided by Richard Isnor when he invited me to take a Visiting Fellowship at Canada's International Development Research Centre (IDRC) in 2008. The book was one of the expected outcomes, and was strongly supported by the then President, Maureen O'Neil. Some months before, Susanne Huttner had been appointed as Director of the Directorate of Science, Technology and Industry (DSTI) of the Organisation for Economic Co-operation and Development (OECD). She invited me to join the management team responsible for the OECD Innovation Strategy to be delivered in 2010. Work on the Innovation Strategy and on the book were mutually reinforcing.

Much of the content of the book derives from working with delegates to the OECD Working Party of National Experts on Science and Technology Indicators (NESTI). These colleagues are too numerous to name but I will thank Giorgio Sirilli for his years as Chair and Alison Young for her support of the working party. I will also thank Erika Rost who unfailingly offered helpful comments on my contributions to the work of NESTI and who has read, and commented on, every word of this book. The work done with NESTI, which started in 1988 could not have happened had Statistics Canada not been supportive. That support was led by the Chief Statistician of Canada, Ivan Fellegi, until he left Statistics Canada in 2008.

Another benefit of NESTI was the opportunity to work with many good people from Eurostat (Statistical office of the European Communities) and with colleagues from observer countries. Leonid Gokhberg and his colleagues from Moscow have provided stimulating exchanges over the years and organized influential conferences.

When I became Chair of NESTI in 2002 I made a point of accepting invitations to talk about the importance of science, technology and innovation indicators and their potential role in policy. These talks reinforced my thoughts about the need for this book and the first invitation was from Liu Shumei of the Ministry of Science and Technology (MOST) in Beijing in 2004. It was followed by an invitation from Mario Albornoz to Sao Paulo to address a Network on Science and Technology Indicators – Ibero-American and Inter-American (RICYT) conference. However the concept of the book came together at a seminar given in Warsaw at the Institute of Economics of the Policy Academy of Sciences in May 2006. This was organized by Tadeusz Baczko and promoted by Grazyna Niedbalska of Central Statistical Office. It was the discussion following the seminar that convinced me that the book should be written, and the reader can find it in *The Future of Science and Technology and Innovation Indicators and the Challenges Implied*, published by the Polish Institute of Economics in 2009 and edited by Tadeusz Baczko. That set the stage for a meeting with Matt Pitman at a conference who suggested that Edward Elgar publishing would be interested in the book. Matt has been most supportive during the period of gestation.

Many of the findings in the text come from years of work at Statistics Canada which would not have been possible without my former colleagues Frances Anderson, Daniel April, Michael Bordt, Louis Marc Ducharme, Louise Earl, Charlene Lonmo, Antoine Rose, Susan Schaan and George Sciadas. Martin Wilk, as Chief Statistician, hired me and then cancelled the project I was hired to run, ensuring a career involving science, technology and innovation indicators, an alternative path that was then available. Bert McInnis and Rob Hoffman brought me into the organization, arguing that the statistical office was the CERN (European Organisation for Nuclear Research) of the social sciences, an interesting metaphor that I continue to ponder.

In the course of developing indicators at Statistics Canada, there were several activities involving international collaboration. One was a series of five workshops to explore the link between indicators and policy. They each involved international colleagues and gave rise to five books edited by John de la Mothe and his co-editors Gilles Paquet, Jorges Niosi, Dominique Foray and Al Link. The topics covered were local and regional innovation, information and communication technology when the definition of electronic commerce was being debated; biotechnology when a statistical programme was being established at the OECD; knowledge management as part of an OECD project; and networks, alliances and partnerships, supported by the National Science Foundation which had a common interest in the subject. A 2003 workshop on innovation was a precursor to the OECD Blue Sky II Forum and produced a book edited by Louise Earl and myself. There were also OECD forums, the OECD High-Level Forum on Knowledge Management in 2000 and the OECD Blue Sky II Forum in 2006, both supported by the OECD and organized by Louise Earl. The Blue Sky II Forum was also supported by the National Science Foundation and Industry Canada.

The work at Statistics Canada benefited greatly from the Advisory Committee on Science and Technology Statistics and its three Chairs,

Stephen Feinberg, Susan McDaniel and Tom Brzustowski. The committee created working parties in the 1990s that helped to define the programme and to secure its financial support. While all of the members of the committee made significant contributions, I would single out Martin Walmsley for his ability to manage complex and profound concepts and to build a consensus around them in demanding circumstances.

The link with the Division of Science Resources Statistics (SRS) of the US National Science Foundation has been mentioned, but I should mention Lynda Carlson as a valuable colleague at the OECD and strong supporter of collaboration between the SRS and Statistics Canada. John Jankowski has always been willing to share knowledge and to offer comments on research projects and on this book. In the area of the Science of Science and Innovation Policy (SciSIP) programme of the NSF, Julia Lane has provided useful input.

The reader will notice that user innovation is a theme throughout the book. This goes back to a survey done by the author in 1987, a survey managed by Robert Tinari at the US Census Bureau in 1988 and the publication in that year of a book by Eric von Hippel on the role of the user in innovation. This eventually led to a collaboration with Eric von Hippel and a paper on user innovation based on a survey run by Susan Schaan and Mark Uhrbach at Statistics Canada. As far as this book is concerned, Eric has read every word on user innovation and has provided comments.

Another recurring theme in the book is innovation in developing countries. This reflects the author's collaboration with the New Partnership for Africa's Development (NEPAD) initiated by John Mugabe in 2005, many discussions with Michael Kahn when he was the South African observer at NESTI, and earlier work in South Africa. Aggrey Ambali and Philippe Mawoko have continued the collaboration with the NEPAD Office of Science and Technology and Philippe Mawoko provided comments upon Chapter 9. It has also been a pleasure to be associated with the Research Policy Institute in Lund, with Claes Brundenius, and the Swedish International Development Cooperation Agency (SIDA)-supported project on innovation and R&D surveys.

As part of the work on the Innovation Strategy, a workshop was organized in Paris in January 2009 on converting knowledge to value in developing countries. This was a joint OECD–UNESCO project with support from IDRC and SIDA. Tony Marjoram from UNESCO (the United Nations Educational, Scientific and Cultural Organization), Gang Zang from the OECD and Jean Woo from the IDRC helped make the workshop happen. Jean Woo also contributed substantively to Chapter 9.

The book has benefitted from discussions with Howard Alper, the Chair of the Canadian Science, Technology and Innovation Council (STIC), and with Patricia Malikail and her team in the STIC Secretariat. There have also been discussions with Peter Nicholson over the years, and more recently in his role as President of the Council of Canadian Academies (CCA). The subject was the work of a CCA panel looking at innovation and productivity in Canada.

At the end of my IDRC Fellowship I was fortunate to be able to join UNU-MERIT (United Nations University Maastricht Economic and Social Research and Training Centre on Innovation and Technology) as a Professional Fellow and have the opportunity to manage IDRC-supported case study projects in African countries. I am also Professor Extraordinaire of the Tshwane University of Technology (TUT) associated with the Institute for Economic Research on Innovation (IERI) at TUT. I am grateful to Luc Soete for the former appointment and to the Tshwane University of Technology for the latter. Rasigan Maharajh and Mario Scerri, of IERI, commented on Chapter 9. In the UNU-MERIT context I should acknowledge the work and contribution of Anthony Arundel.

The debt owed to colleagues at the OECD over the years is large. During my work on the Innovation Strategy I have had helpful discussions with Pier Carlo Padoan, Richard Carey, John Dryden, Enrico Giovannini, Susanne Huttner, Daniel Malkin, Michael Oborne, Nabuo Tanaka and Andy Wyckoff. Various colleagues are acknowledged in the text where they have made specific contributions. In the indicators domain, I have had a long and fruitful working relationship with Alessandra Colecchia. As the Innovation Strategy moves forward, I am working with Miraim Koreen.

A final acknowledgement goes to Paris, where the book was given structure and partly written. In the early twentieth century the existentialist debate took place in the 6th arrondissement at Les Deux Maggots and the Café de Flore. OECD delegates tend to stay in the 16th arrondissement and the early debates, in the late twentieth century, some heated and long, took place in Le Marty on rue de Passy. As part of creative destruction, Le Marty closed, and the debate moved to Le Passy where it continues. Outside of the OECD, it is the only place where I have been able to hold a discussion of the four components of the Oslo Manual definition of innovation.

While there have been many inputs and useful comments, the final text and any errors are the responsibility of the author.

Abbreviations

AAAS	American Association for the Advancement of Science
AEC	African Economic Community
AMCOST	African Ministerial Council/Conference on Science and Technology
APEC	Asia-Pacific Economic Cooperation
ARPA-E	Advanced Research Projects Agency – Energy
ASEAN	Association of South East Asian Nations
ASTII	African Science, Technology and Innovation Indicators
AU	African Union
BEA	Bureau of Economic Analysis (US)
BERR	Department for Business, Enterprise and Regulatory Reform
BIAC	Business and Industry Advisory Committee (OECD)
BIS	Department for Business, Innovation and Skills (UK)
BMBF	Federal Ministry of Education and Research (Germany)
BRDIS	Business R&D and Innovation Survey (US)
CCA	Council of Canadian Academies
CEC	Commission of the European Communities
CERN	European Organization for Nuclear Research
CIP	Competitiveness and Innovation Framework Programme (CEC)
CIS	Community Innovation Survey
CPA	Consolidated Plan of Action (NEPAD)
CSTP	Committee for Scientific and Technological Policy (OECD)
CYTED	Latin-American Science and Technology Development Programme
DAC	Development Assistance Committee (OECD)
DEIP	Design and Evaluation of Innovation Policy in Developing Countries (UN)
DFID	Department for International Development (UK)
DG	Directorate-General
DIUS	Department for Innovation, Universities and Skills (UK)
DSTI	Directorate of Science, Technology and Industry (OECD)

DUI	learning by doing, using and interacting (Lundvall)
EIT	European Institute of Innovation and Technology
EIS	European Innovation Scoreboard
ESA	European Space Agency
ESFRI	European Strategy Forum on Research Infrastructures
EU	European Union
Eurostat	Statistical Office of the European Communities
GDP	gross domestic product
GERD	gross domestic expenditure on research and development
GM	genetically modified
GMES	Global Monitoring for Environment and Security (CEC)
HAP	Heiligendamm-L'Aquila Process
HCST	Haut Conseil de la Science et de la Technologie (France)
HRDC	Department of Human Resources Development Canada
HRST	human resources in science and technology
ICT	information and communication technology
IDRC	International Development Research Centre (Canada)
IERI	Institute for Economic Research on Innovation
IP	intellectual property
IPR	intellectual property rights
ISO	International Organization for Standardization
ISTWG	Industrial Science and Technology Working Group (APEC)
ITER	International Thermonuclear Experimental Reactor
JTI	Joint Technology Initiative (CEC)
KICs	Knowledge and Innovation Communities (EIT)
KIS-IP	Knowledge-Intensive Services Innovation Platform (CEC)
LEED	Local Economic and Employment Development (OECD)
LMI	Lead Market Initiative (CEC)
LTG	Limits to Growth
MDGs	Millennium Development Goals
MEBSS	materials/energy balance statistical system
MFP	multifactor productivity
MIP	Mannheim Innovation Panel (ZEW)
MNEs	multinational enterprises
MOST	Ministry of Science and Technology (China)
NAICS	North American Industry Classification System
NEPAD	New Partnership for Africa's Development

NESTI	Working Party of National Experts on Science and Technology Indicators (NESTI)
NGO	non-governmental organization
NRC	National Research Council (US)
NRC-IRAP	National Research Council Industrial Research Assistance Program (Canada)
NSF	National Science Foundation (US)
NSTC	National Science and Technology Council (US)
OAS	Organization of American States
OAU	Organization of African Unity
OECD	Organisation for Economic Co-operation and Development
OST	Observatoire des Sciences et des Techniques (France)
OST	Office of Science and Technology (NEPAD)
OSTP	Office of Science and Technology Policy (US)
PCAST	President's Council of Advisors on Science and Technology (US)
PRSPs	Poverty Reduction Strategy Papers
R&D	research and development
RICYT	Network on Science and Technology Indicators – Ibero-American and Inter-American
S&T	science and technology
SAREC	Swedish Agency for Research Cooperation
SBIR	Small Business Innovation Research (US)
SBRI	Small Business Research Initiative (UK)
SciSIP	Science of Science and Innovation Policy
SERF	Socio-Economic Resource Framework
SIDA	Swedish International Development Cooperation Agency
SMEs	small and medium-sized enterprises
SNA	System of National Accounts
SoSP	Science of Science Policy
SR&ED	Scientific Research and Experimental Development
SRS	Division of Science Resources Statistics (US)
STI	science, technology and innovation
STIC	Science, Technology and Innovation Council (Canada)
TEP	Technology Economy Programme (OECD)
TPP	technological product and process (innovation)
TUAC	Trade Union Advisory Committee (OECD)
TUT	Tshwane University of Technology
UIS	UNESCO Institute for Statistics
UN	United Nations

UNCTAD	United Nations Conference on Trade and Development
UNESCO	United Nations Educational, Scientific and Cultural Organization
UNIDO	United Nations Industrial Development Organization
UNU-MERIT	United Nations University Maastricht Economic and Social Research and Training Centre on Innovation and Technology (The Netherlands)
ZEW	Centre for European Economic Research (Germany)

PART I

Issues and frameworks

1.   A challenging world

INTRODUCTION

In 2009 when this book was written, the economy was in recession and people were asking what went wrong. Part of what went wrong was innovation in financial services which resulted in the release of attractive new products to the market. They diffused rapidly and widely and then lost value. The rest is history, a painful history for those who lost homes, savings and businesses. As the first signs of recovery appear, the question being asked is whether this can happen again.

This book looks at innovation, what it is, how it is measured, and how policies are developed and implemented to support it. In doing this, the framework conditions are examined, like market regulation, and the costs of doing business, to see how the present situation could be avoided. However, framework conditions are not just the work of government, as culture and history contribute. The stigma, in some countries, of bankruptcy is an example which reduces the likelihood of taking risk.

While innovation has been around since markets began, understanding it and the policies that support it remains a challenge. Once the issues are reviewed, consideration is given to a research agenda for those people who create innovation policy, implement it, measure the activity of innovation in the economy, and provide the statistics and indicators which are used for monitoring and evaluating the effects of policy intervention. The better understanding of innovation and innovation policy may result in better economic and social outcomes from these activities.

Part I sets the stage for a discussion of innovation, its measurement and the use of the resulting indicators as part of the policy process in Part II. Chapter 1 makes the point that the discussion of innovation and innovation policy takes place in a global, complex, dynamic and non-linear system, a phrase that recurs throughout the book, and at a time of world financial crisis. The goal for most countries is to manage sustainable productivity growth, driven by innovation, supported by effective policy. Chapter 1 also lays out a set of stylized facts about the institutions that contribute to the activity of innovation to help structure questions that the reader should be asking while reading the rest of the book. Chapter 2 presents systems frameworks that organize many of the topics that have to be considered in a discussion of innovation. Armed with the motivation for learning about innovation, especially in the present climate of rapid economic and social change, the stylized facts to support pointed questions, and the systems frameworks, the reader should be ready for the following chapters.

Innovation and Sustainable Productivity Growth

This book is about innovation strategies for a global economy, their development, implementation, measurement and management. It was written in 2009 when there were economic and social impacts resulting from a global financial crisis which can, itself, be linked to innovation in financial services. It is also a time of global challenges following from changes in climate, the supply of energy, food and water, and the return of infectious diseases that were thought to be under control.

Meanwhile, people in the industrialized countries, who are expected to deal with these problems, are ageing. They control the wealth and the knowledge needed to effect real change. As they age, knowledge is being lost and governments are struggling, not just with the cost of the present financial crisis, but with the growing cost of caring for their ageing populations.

It is a time for action on the part of governments when there are progressively fewer productive people to support the action and pay for the increasing demand for social services. Innovation, the creation of value from knowledge, and a driver of economic growth if it is well managed, is seen as a way forward, but subject to a number of constraints. With fewer people producing the knowledge needed to create value, innovation has to deliver increased productivity leading to economic growth. In view of the global challenges of climate change, and the limited supply of energy, food and water, the growth resulting from innovation must be sustainable. The goal is sustainable productivity growth, driven by innovation.

The question for governments is how to promote innovation; a challenge in itself, and made greater by the need for the innovation to result in sustainable productivity growth. This in a global economy, at a time of rapid change, when it is clear that innovation does not happen in isolation but in a global, complex and dynamic system, that is non-linear in its response to policy intervention. Non-linearity here simply means that a new policy intervention may not result in an expected outcome because of the feedback loops in the system that link it to other policy interventions, and framework conditions, in ways that are difficult to predict.

The question for firms is how they engage with the other actors in the innovation system to ensure that the framework conditions that result from culture, history and government regulation are supportive of innovation that will contribute to the goals of the society. Firms know how to innovate.

Understanding a global, complex, dynamic and non-linear innovation system sufficiently to support policy interventions that result in the realization of national goals is a non-trivial problem and one that this book is not about to solve. The book is divided into four Parts, and Part I discusses the issues and frameworks needed to present the problem. Part II provides a language of discourse and examines the role of statistical measurement, and the use of the resulting indicators, needed to inform the discussion. Part III then looks at what is being done with innovation strategies at a critical time in the world economy, and gives some tools for their development. Part IV extends the discussion of innovation strategies to include their application in developing countries, and adds new topics to the discussion to be addressed in the medium term, and some topics that need to be addressed in the short term. Then, a work programme is offered to the people that develop and implement innovation strategies.

To continue with Part I, the intention is to provide a historical context for the present situation, raise some questions about understanding the innovation system – some of which can be addressed with existing knowledge – and then to lay out the basis for a discussion intended to lead to more work and a deeper knowledge of the system and how to influence its behaviour.

Recent History Leading to Present Needs

The world was a different place in 1989. The Berlin Wall was about to fall, to reveal economic and social opportunities in Central and Eastern European countries; the mobile telephone had yet to dominate the way people lived and did business; and the growth of Brazil, China and India had not yet dominated the economic debate. Tom Friedman (2006) has reviewed this period in depth, as well as the issues around climate change (Friedman 2008). Here only highlights are selected to set the stage for a discussion of innovation and innovation strategies.

Since 1989, there have been many years of growth in the Organization for Economic Co-operation and Development (OECD) countries, with some slowdowns, a spreading of information and communication technology (ICT) infrastructure which slowed or stopped after the dot com meltdown which began in November 2000, a rise and fall in the price of oil, and a change in the eating habits of people in the emerging economies as the newly affluent moved from grains to meat as a principal part of their diet.

Climate change became more of an issue, and biofuels were introduced to reduce dependence on imported oil and, in principle, to reduce greenhouse gas emissions. As the first generation of biofuels came from edible plants, the coupling of energy and food policies, along with the increased cost of transporting food, natural disasters, and the change in eating habits, resulted in food shortages which were thought to be a thing of the past. Famine had returned.

The tragic events of 11 September 2001 transformed geopolitics and made security a pervasive priority. With global warming and civil unrest in parts of Africa, diseases that were once thought to be under control began to return, helped by the prevalence of HIV/AIDS. World consumption of water increased and sources vanished with increasing temperature or were degraded by industrial processes or human or animal pollution. Supply of fresh water became an issue, and not just in developing countries. War, pestilence and death continued as matters of public concern.

Climate change, energy and food costs, and water supply are global challenges in the physical world which affect human health, poverty and equity, and security. Reviewers of disasters of the past, such as Diamond (1997) and Wright (2004), would argue that much of what the world is now experiencing has already been seen but on a smaller scale. The difference now is that the scale is global and the consequences of a global failure would be immense and could be terminal. Avoiding such failure is why the global challenges must be the principal motivators of better innovation leading to sustainable productivity growth.

There are challenges in the economic world that arise from the inability of the regulatory systems of the industrialized countries to deal with financial issues, an illustration of which is the subprime crisis. The financial problems led to reduction in economic growth and contributed to the other challenges faced by humanity in 2009. This example makes the point that innovation does not always lead to growth; it can damage the global economy and the society, and that is a motivation for producing innovation strategies that in the worst case do no harm, and in the best case result in the outcomes desired.

The period since 1989 has been one of rapid change, new products and processes, practices, and markets, have evolved and are changing the way business is done. Natural resource, manufacturing and service industries are no longer local, but global, and there is a challenge in managing global value chains that cross many jurisdictions. There has always been innovation in response to the opportunities provided by change but the changes of the last two decades have been unprecedented in the speed with which they have occurred. This poses another challenge, and that is to understand how innovation actually works, locally and globally, and how it changes with time. Knowledge of the process of innovation can contribute to well-being and help people, and their governments, to address global challenges.

However, understanding innovation and the dynamics of change, and harnessing that knowledge, is not simple and it is not clear that the experience of the past provides guidance for the future. John Marburger, who was the science advisor to US President G.W. Bush, pointed out that: 'in the face of rapid global change, old correlations do not have predictive value' (Marburger 2007). This means that to address the challenge more has to be learned about how innovation happens. Chris Freeman and Luc Soete (2007) have suggested that: 'the link between the measurement of national STI [science, technology and innovation] activities and their national economic impact, while always subject to debate, particularly in the context of small countries, has now become so loose that national STI indicators are in danger of no longer providing relevant economic policy insights'. Both observations pose challenges to those trying to understand the activity of innovation sufficiently to inform the policy process. Both were made at the OECD Blue Sky II Forum in 2006 which was held to develop new indicators and measurement methods to improve the understanding of innovation (OECD 2007a).

Under the Swedish European Union Council Presidency, the Lund Declaration[1] was issued following a conference of 400 researchers and politicians in Lund, as a response to the challenges just discussed, and it opens with the statement that: 'Europe must focus on the grand challenges of our time'. The Lund Declaration is about research, not about innovation, but the proposals are also relevant to innovation. The three leading points of the Declaration are the following:

- European research must focus on the Grand Challenges of our time moving beyond current rigid thematic approaches. This calls for a new deal among European institutions and Member States, in which European and national instruments are well aligned and cooperation builds on transparency and trust.
- Identifying and responding to Grand Challenges should involve stakeholders from both public and private sectors in transparent processes taking into account the global dimension.
- The Lund conference has started a new phase in a process on how to respond to the Grand Challenges. It calls upon the Council and the European Parliament to take this process forward in partnership with the Commission.

These points make clear the importance of alignment of policy instruments, the need for involvement of stakeholders from the private and public sectors, and the start of a new phase in responding to the 'Grand

Challenges'. Policy alignment, and high-level dialogue, will recur in subsequent chapters of the book. To discuss the problems of policy alignment, whether innovation policy or research policy, in a complex system, a systems approach is needed.

A Systems Approach

The systems approach is ideal for understanding the activities of the actors in the system, their linkages and the outcomes of their activities and linkages, leading to economic and social impacts. It is a means of classifying the components and also is a basis for thinking about dynamics and feedback loops. In Chapter 2, the systems approach is presented as an analytical tool, not as the basis for a theory of innovation.

In Chapter 2, another system is presented which predates this discussion of innovation by decades and which is used for presenting a macro view of the economy (but not of the welfare of the society), the System of National Accounts (SNA). The SNA is part of the discussion here as it has an established set of definitions and practices, codified in manuals and used throughout the world to produce internationally comparable data for indicators such as gross domestic product (GDP), inflation, employment, trade and investment. The SNA indicators support evidence-based fiscal and monetary policy and are used by finance departments and central banks; by economists, generally in their research; and in econometric models on which economic predictions are based. The innovation community has a long way to go before the work of the government departments of education, research, technology or training can be informed by such a well-established set of indicators, and a community of scholars able to work with them to draw inferences and provide policy advice. However, indicators, models and informed advice are not the ultimate goal for the innovation community, given that the current financial crisis happened with all of that machinery in place. The goal is to address the global challenges, to improve social welfare, and to avoid getting it wrong.

Language and Learning

In Part II, the definitions of innovation, and related terms, are presented so that an innovation system can be discussed in an unambiguous manner. The development of a language of discourse simplifies the exchange of ideas, but first the language has to be developed, diffused broadly, learned and used.

The place where the concepts and definitions of innovation and the interpretation of data on innovation evolved and were codified was at the OECD in the Working Party of National Experts on Science and Technology Indicators (NESTI). The story is told in greater detail in Chapter 3, but the point to be stressed here is the evolution, since the start of discussions of how to measure innovation in the late 1980s, of a language to describe innovation and definitions which supported the collection of data through surveys, from administrative sources and from case studies. The Oslo Manual, which was originally an OECD manual, became with the second edition the joint responsibility of Eurostat, the European Union (EU) statistical office, and the OECD. As a result the manual, and the language, are used in all EU and OECD countries as well as in some observer countries that participate in the work of these organizations.

Language is not static, and as the activity of innovation was probed through innovation surveys, of which the Community Innovation Survey (CIS) was the main example, the language, and its domain of application evolved. With the third edition of the Oslo Manual (OECD/Eurostat 2005), the definition had four components, adding organization and business practices, and market development, to the original two which were product and process innovation. As well, non-technological innovation was added to technological innovation. The domain of application extended to the entire market economy, having started with just manufacturing in the 1990s. The survey activities, which use the concepts and definitions of the Oslo Manual (OECD/Eurostat 2005), are described in Chapter 4.

The evolution of the language was a group learning exercise, informed by the survey results discussed at NESTI and Eurostat meetings and by the solving of technical problems. The language was also influenced by the need to be able to inform the policy process at the end of the day. The hope was also that policy learning would keep pace with the development of the surveys and indicators, and that there would be a recognition that innovation policy goes beyond research and development policy. The use of indicators in the policy process is discussed in Chapter 5 where the point is made that the broader discourse, and learning process, has got a long way to go before innovation is a topic for popular discussion.

Innovation Strategies, Components and Coordination

In Chapter 6, there is a review of what is going on to respond to the need for more effective innovation strategies and the plans of the Commission of the European Communities (CEC) and of the OECD are reviewed. As the work of both the CEC and of the OECD are 'works in progress' no definitive comments can be made, but the two processes, which are quite different, can be compared. The *Strategy for American Innovation* (Executive Office of the President 2009), released in September 2009, is mentioned briefly.

The discussion of the work of the CEC and of the OECD leads to a review in Chapter 7 of what a possible set of components of an innovation strategy could be. Then, in Chapter 8, the question is how these components, or a subset of them, could be coordinated by government(s) to deliver an effective strategy or strategies. These observations are based on country experience, not on a theoretical position. If there is a single finding that emerges, it is that there is no single solution to the creation of an innovation strategy.

Broadening the Horizons and Next Steps

Part IV goes beyond the discussion of what is happening now in developed economies and addresses the place of innovation strategies in developing countries, as they are part of the global economy and a vulnerable part of the global society. In addition, a series of topics are examined which take the discussion beyond what is now being done in producing and applying innovation strategies. This includes public sector innovation, already evident in the response to the financial crisis, and the need for a new social science, the science of innovation policy.

The discussion of the science of innovation policy is based on the call by Marburger for a Science of Science and Innovation Policy (SciSIP) which is now a programme of the US National Science Foundation. As 'science' is quite different from 'innovation', Chapter 10 focuses on the science of innovation policy and argues that this work is essential if there is to be an understanding of a global, complex, dynamic and non-linear innovation system sufficient to support policy learning and effective intervention in the medium term. This understanding is also necessary to give ministers responsible for aspects of innovation better advice than has been received by finance ministers and central bankers for decades. This is a significant goal, as ministers of innovation are there to resolve the global and local challenges while ministries of finance and central banks are there to provide the fiscal and monetary stability to support the work needed to achieve the goal.

What should be done in the medium and the short term provides an agenda for future work in Chapter 10. Chapter 11 turns to the people who deal with innovation policy, analysis and measurement, and offers them an agenda intended to bring a coherent approach to work on the subject of innovation policy and its understanding.

A reader with an interest in policy can go directly to Parts III and IV, but all of the discussion of the evidence needed for evidence-based policy is in Part II, and part of the underlying view in the text is that the communities of policy-makers and indicator producers should overlap and learn from one another to avoid the waste of survey results that are never used, or policy targets that can never be achieved within existing socio-economic and physical boundaries. Earlier thoughts on these issues may be found in Baczko (2009), Earl and Gault (2006), in Chinese in Gault (2004), and in Spanish in Gault (2008a). Gault (2009) provides a view of innovation strategies in Russian.

SOME STYLIZED FACTS ABOUT INSTITUTIONS AND INDIVIDUALS

The reader is encouraged to reflect on some findings from work on innovation measurement and policy that should be well known to the experienced practitioner, but might not be to the policy analyst or survey statistician embarking anew on this subject. These facts should be kept in mind when reading this text and are useful for asking probing questions which this text, or others, may not be able to answer. Critical questioning of the concepts, definitions and evidence is part of language development, learning and change. In what follows, the focus is on the firm, but education, government and institutions are mentioned.

Firms

One place where innovation happens is in firms
Firms produce a new product (good or a service) and put it on the market. It introduces a new production or delivery process and it is the firm that changes its organization or adopts new business practices or develops new markets. In what follows, process innovation includes any or all of the activities in the previous sentence. Individuals produce product innovation and there is work on public sector innovation. The point is that innovation happens in firms, but this is not the only place where innovation happens.

Firms are connected to other institutions
Firms buy goods and services, hire people, convert inputs to outputs and sell the results to other institutions, including firms and households. Governments collect taxes from and provide incentives to firms. They also set and enforce rules. Universities and research institutions provide highly qualified people and knowledge. The education sector provides the trained people who make up most of the labour force. Describing the actors, the

activities, the linkages and the outcomes of the activities and linkages is why the systems approach in Chapter 2 is needed.

Firms are constrained

Firms operate in a multidimensional box, the walls of which are formed by available inputs, culture, infrastructure, location, history and regulation. Some simple examples are the inability of a firm to double its research and development (R&D) activity if the researchers are not there to be hired, or to sell its genetically modified food products to countries that regulate against their imports. Other examples will be developed in the text. Modelling of the system to include constraints, both physical and institutional, is discussed in Chapters 2 and 10.

Most firms are SMEs

As most firms are small or medium-sized enterprises (SMEs), size is a useful analytical variable. For understanding firm behaviour, the reader as analyst should distinguish between firms with zero employees, such as academics acting as consultants and nannies selling their services to house-holds, and firms with employees. In Canada, there are about 1 million firms with employees and another million without employees. Put them together and a skewed distribution becomes even more skewed.

For a total population of firms with employees, expect 98 per cent to have fewer than 100 employees, 95 per cent to have fewer than 50, and 88 per cent fewer than 20. These are approximate figures and will vary by country and by sector, but the reader should know what they are for their own country so that they can ask questions about firm-level analysis.

Readers with access to counts of firms and employment for their country may wish to try fitting the following equation to the data:

$$N(E) = N_1 E^{-\alpha}$$

where N is the number of firms with E employees, N_1 is the number of firms with 1 employee (a large number), $E \geq 1$ and α is a constant. This is a power law and it provides a useful way of storing the information about numbers of firms and employees.[2]

Large firms are complex

By the time a firm becomes large (250 employees or more is one definition, 500 or more is another), it may have more than one place where it does its business, it may engage in more than one industrial activity and it may have activities outside of the country. If it operates in several countries it is a multinational enterprise (MNE). The terms used to describe a firm, and its components, change with the industrial classification used in the country. This will be elaborated upon in Chapter 4, but the fact to retain is that large firms, with large turnover,[3] active in several industries in several countries, are complex entities, difficult to understand and they can distort statistical analyses if they are not treated correctly. It is for this reason that some indicator reports use only data on SMEs. The results may ignore about half of the total turnover (revenue) in the industry, but the statistics are simple.

More firms innovate than do R&D

The difference between the propensity to innovate and the propensity to do R&D is size dependent. Large firms, on the whole, innovate and do R&D. For medium and small-sized firms the difference is quite significant and has implications for innovation strategies such as those that deal with the creative destruction of the Schumpeter Mark I regime (SMEs) and the creative accumulation of knowledge of the Schumpeter Mark II regime (large incumbents), and the balance between the two. This is discussed in Chapter 8.

Business R&D is concentrated

There are a few firms that do most of the R&D in a country. In Canada, 75 firms do half the R&D, and about 19 000 firms do the other half. This has implications for how R&D policy, as part of innovation policy, could be done.

Space matters

A firm in a major city has access to more resources – human, infrastructure and material – than firms in more isolated regions. This is reflected in the innovation cluster literature.

Time matters

The lifetime of firms varies from industry to industry and the volatility resulting from the birth and death rate of firms is a factor in innovation. The time required to start a firm or to take it into bankruptcy are also factors. The speed with which a firm can respond to market opportunities is a key factor in innovation as it reflects the ability of the firm to learn, to change as a result of learning, and to gain market share.

Learning matters

Firms learn differently in different circumstances and learning is size dependent. As discussed in Chapter 5, there are discontinuities in knowledge management practices used by firms as they increase in size and such discontinuities do not just apply to knowledge management. Where these discontinuities occur is critical to how the firm is able to innovate.

How firms innovate matters

Firms can be producer innovators or process innovators, or both, and recall that 'process' here includes firm organization, management practices and market development as well as production and delivery of the product.

Product innovating firms put new goods or services onto the market. The information needed for the innovation may originate entirely within the producer firm. This is producer innovation. However, once the product is sold to a customer, information flows can occur through service agreements, visits from sales and marketing staff, or by customer-initiated exchange such as complaints and requests for upgrades or for related goods or services. In every innovation survey users or customers rank high on the list of sources of information for innovation. In some circles this is referred to as 'user-driven innovation'. User knowledge flows to the producer and better products result, 'driven' by the user.

When users provide prototypes or complete sets of plans for the producer, and the producer bases a new or significantly improved product substantially on these inputs, then users, and not producers, are the innovators. In that case the appropriate term is 'user innovation' (product). Types of product innovation are summarized in Table 1.1.

Process innovation is different. A process is changed to solve a problem in order to get a product to market. It is an activity that is internal to the firm, but the process technologies or practices could be products sold by a producer firm. In fact, in most cases they are. The types of innovation are summarized in Table 1.2.

What is different in the case of user innovation is what user innovators do with the intellectual property (IP) that they create when they modify or develop a process. There is evidence, discussed in Chapter 5, that user-innovators in some cases give away the intellectual property, and this has implications for intellectual property policy as part of innovation policy. Producer innovators have a higher propensity to protect intellectual property using conventional IP instruments.

Education Sector

The education sector will appear throughout the book, but from the perspective of stylized facts the points to retain are that the education sector produces most of the labour force as well as the highly qualified personnel, but both groups make significant contributions to innovation.

Table 1.1   Product innovation definitions

Inputs for product innovation	Type of innovation
Producer only:	
Information from producer	Producer innovation (product)
Producer and user:	
Information from user	User-driven product innovation
Prototype or blueprint from user	User innovation (product)

Table 1.2   Process innovation definitions

Inputs for process innovation	Type of innovation
User buys a process innovation- enabling product from producer	

User uses the product to create an in-
   house process innovation
	Producer-driven process
	   innovation*

User modifies the product and uses it to
   create an in-house process innovation
	User innovation (process):
	   Modifier

User develops a process innovation-
   enabling product and uses it to create an
   in-house process innovation
	User innovation (process):
	   Developer

Note: * The product may or may not be an innovation for the producer. It can be an innovation for the user if it is new to the firm and is introduced in the reference period of the survey that is measuring the activity of innovation.

The education sector also creates knowledge which can be made freely available through publication, or it can be commercialized and sold. Knowledge also flows through graduates who embody knowledge and through consultancy which may be undertaken by staff.

Government Sector

The government sector is responsible for building and maintaining some of the boundaries of the innovation system. It cannot deal with all of the boundaries as, in a global economy, some cannot be controlled by single governments. Regulating global financial markets is an example. The government controls the flow of people through immigration policy, and acts directly to support innovation through grants, contracts or contributions, or indirectly through tax incentives and regulations. It also is the source of new policies and their implementation, and that makes the government sector key to the discussion.

Governments, and other public institutions, can engage in innovation activities, but this discussion is left to the section below on 'Other forms of innovation' and to Chapter 10.

Individuals

Individuals are sources of innovation. Their views on consumer products are sought by producers as part of improving their product. In this role, the consumer is involved in user-driven innovation (see Table 1.1). Individuals are also innovators. They can change a product to suit their needs or create a new product.

The question, then, is what they do with the knowledge gained as a result. If they present it to a producer as a prototype or a blueprint, this is a case of user innovation (Table 1.1). However, they may share their innovation among the members of a community of practice. This takes the activity beyond the bounds of innovation as there is no conventional market. This is a similar conceptual issue to that found in dealing with public sector innovation, and it is discussed again, along with public sector innovation, in Chapter 10.

The point to retain is that users (consumers) play key roles in innovation. They contribute to user-driven innovation and user innovation.

OTHER FORMS OF INNOVATION

User Innovation

User innovation in firms is seen in the statistics for process innovation. There is also user innovation in respect of products. This should be distinguished from user-driven innovation which recognizes the user as source of information which is a key input to producer innovation.

End users may bring their product innovation back to the producer in the hope of having a better product produced, or they might share the knowledge of how to create it among a community of practice (surgeons developing new operating tools, for example). The question is: how large is this activity and what are the implications for innovation policy? This is considered further in Chapter 10.

Public Sector Innovation

Most of the text deals with innovation in firms and the problems of getting products to market. However, all of the activities that contribute to innovation, such as R&D, capital investment, training and development, and acquisition of intellectual property can be, and are, done in the public sector. The public sector in a number of countries acted quickly, and in unconventional ways, to resolve some of the problems being caused by the financial crisis. There is no question that the public sector can engage in innovation activities as presented in Chapter 3. The question is whether it can formally engage in the activity of innovation. This question is being addressed in countries that want to see public sector reform, with new or improved products or new or improved processes being used, with better outcomes for the users of public goods and services and better use of public resources. As this is a relatively new area in the innovation business, it is discussed in Chapter 10 and the question is how the two streams of private sector and public sector innovation will be brought together, and the knowledge codified and used to inform future policy.

Social Innovation

The OECD LEED Forum on Social Innovations (OECD 2008a) looks at 'social innovation' in the context of local economic and employment development (LEED) but this has not led to a manual with definitions of social innovation or guidelines for the collection and interpretation of data, due in part to the application dependence of the activities and the difficulty of distilling a common set of practices, and measures of linkages, outcomes and impacts. The distinction between economic innovation, which is the province of the Oslo Manual, and social innovation, is that the 'latter deals with improving the welfare of individuals and community through employment, consumption or participation, its expressed purpose being to provide solutions for individual and community problems' (OECD 2008a). The community focus of the work on social innovation could support analysis of social impacts of innovation in communities and this could have direct application in work in developing countries. The interested reader is referred to the reference provided, as the subject of social innovation is beyond the scope of this book.

DEVELOPMENT

Virtually all of this text can be applied in developing countries, as the emphasis throughout is on innovation based on knowledge from any source and not just on knowledge formally created through R&D. However the contextual issues are different. The global challenges are having a greater impact in developing countries than in the developed countries, at least up to now. Urbanization is changing the development landscape as more people live and work in cities, but agriculture is still a driving force in development, especially when viewed as a knowledge-based industry.

Getting innovation strategies right in developing countries is critical as innovation is path-dependent and the wrong path can result in significant outcomes, such as the different positions of the economies of Argentina and the US after more than a century of divergence. Getting innovation strategies right requires governments and other public institutions to develop the capacity to learn and use the language needed to talk about innovation and about innovation strategies, and to develop the capacity to implement the strategies. The need for these capacities is not peculiar to developing countries but acquiring them is more urgent.

SUMMARY

This book is about innovation, the language used to discuss it, the strategies used to promote it, and about some of the areas into which the discussion should be going, such as innovation in developing countries and public sector innovation. The focus, for most of the book, is on how firms can be supported in creating value and bringing it to market. As a result, there is little discussion of science policy or of research policy. Innovation is about markets.

It is worth making the point that R&D is not innovation until it connects to the market, and neither is patenting nor publication. This immediately introduces size as an important variable for understanding innovation as it is the large firm that is more likely to do R&D, to patent (in some industries, as not all use patents), to publish and to innovate. The study of large firms is quite different from that of the 98 per cent, or so of firms that are SMEs, and for these firms, more innovate than do R&D, and the challenge is to support more successful innovation and growth in firm size leading to the undertaking of R&D as a natural outcome of an innovation strategy. This is quite different from promoting R&D and then asking why more of it is not commercialized.

A recurring theme in the text is the role of the user in innovation. Eric von Hippel (1988, 2005) has been studying this for decades outside of the mainstream of the Oslo Manual discourse and it is clear that consumers (end users) are not just sources of information for innovation but they are also innovators who change their products to suit their needs. The same occurs for process innovators, as some users modify their technologies to make them do what is needed better and others develop process technologies in the absence of a solution to their problem being available on the market. What user innovators do with their intellectual property raises some questions for intellectual property policy as part of innovation policy.

Chapter 10 offers some concrete suggestions for work to be done in the medium and the short term, and the principal suggestion is that innovation strategy, as opposed to science strategy, should be the focus of high-level, multisector and ongoing discussion with a view to the promotion of innovation and sustainable productivity growth in response to global challenges.

Chapter 11 assigns some further work to be done by the reader. Appendix A provides websites and directions to support further reading beyond what is cited in the References. For the reader with access to microdata, or with resources to commission work by someone who has, Appendix B presents some data projects, most of which could be repeated with current data. The results would contribute to discussions of innovation, innovation policy and the science of innovation policy.

NOTES

1. The full text of the Lund Declaration can be found at: http://www.se2009.eu/polopoly_fs/1.8460!menu/standard/file/lund_declaration_final_version_9_july.pdf.
2. The power law, or combinations of power laws, turn up in many statistical descriptions. See Florida (2002) on Zipf's Law and the relationship between the number of cities and the size of cities; see also AAAS (2009).
3. 'Turnover' is used in Europe and there is some discussion about what to use in other countries. At Statistics Canada 'revenue' is used in innovation surveys. Operating revenue would make it clear that only revenue from operating the business was expected, as total revenue could include interest and dividends. The term 'sales' has been used in some surveys, where the objective has been to identify the value of sales that can be attributed to a new or significantly improved product. In the text, the term 'turnover' will be used.

2. A systems approach

INTRODUCTION

Chapter 1 raised issues that motivate the need for a better understanding of innovation, of innovation policy and of the use of statistical indicators in support of these activities. The challenge is dealing with, if not understanding, a global, complex, dynamic and non-linear system. This chapter lays out a systems approach to this with two objectives. The first is to provide a means of classifying the phenomena that are driving the issues discussed in the text. The second is to provide a basis for discussing dynamics and the importance of modelling the systems as a step towards understanding dynamics, and using this understanding to encourage policy learning through scenario analysis.

A systems approach to economic and social systems has been part of the economic literature for a long time. Herbert Simon (1996) and Jay Forrester (1971) applied systems theory and dynamic analysis to many problems and shaped the thinking of generations of graduate students. Forrester used a dynamic systems model to support the work of the Club of Rome project, Limits to Growth (Meadows et al. 2004), which gave rise to subsequent systems analysis and policy debate.

As models evolved, and more data were required to populate the variables in the models, attention was given to imposing physical constraints on the models so that they could not produce scenarios that required the consumption of more natural or human resources than were physically available. This was the subject of a UN Statistical Office project (Ayres 1978) and projects elsewhere (Gault et al. 1987).

At about the same time the innovation systems approach was developing (Lundvall 1992; Freeman 1987; Nelson 1987, 1988, 1993; Nelson and Winter 1982; Porter 1990), giving rise to the literature on national systems of innovation, regional systems of innovation and innovation clusters, or local systems of innovation. Fagerberg et al. (2004) provide a review of the approaches; an earlier view is found in Dodgson and Rothwell (1994) and a new handbook of innovation is anticipated in 2010 (Hall and Rosenberg 2010).

In the current situation, all of the world economies are linked by flows of goods, services, capital and people. In the European Union (EU), there is freedom of movement of these four things and there is a call for a 'fifth freedom', the free movement of knowledge. The generation, transmission and use of knowledge are all part of innovation and key to the functioning of innovation systems and to the effectiveness of innovation policies. For these reasons knowledge is incorporated into the framework presented in this chapter.

The systems approach can also be applied to the understanding of how innovation policy works, with the potential for supporting the development of a science of innovation policy. This is addressed briefly here and again in Chapter 10.

All of the examples so far have been at the micro level of the actors in the system, such as individual firms. For more years than there have been official statistics on innovation, there has been a system for dealing with macro indicators for managing the economy, the System of National Accounts (SNA) (CEC et al. 1994). The SNA 93 has been revised to become the 2008 SNA. Part of the revision was a change in the way research and development (R&D) is treated and this gives rise to a brief discussion of the SNA and of the role of intangible investments in understanding innovation.

SYSTEMS

What a System Is

A system consists of actors or economic agents. The actors engage in activities, and have linkages with other actors. As a result of the activities and linkages, there are short-term outcomes and longer-term economic and social impacts.

Actors, for the purposes of this discussion, are governments, businesses (including single entrepreneurs), institutions of education and of research, and private non-profit institutions. The activities of the actors are not limited as, for example, health or education activities could be engaged in by any of the actors just listed.

As the focus of the discussion is innovation, a selection of activities of interest could be R&D, invention, innovation, training and development, capital investment and intellectual property protection. More activities are introduced in Part II of the book when the concepts and definitions needed to measure innovation are discussed. Linkages include any interaction between the actors such as contracts; licensing of intellectual property; flows of data, information or knowledge from or to public or private sources, collaboration; and exchange of human resources.

Outcomes can be any consequence of the activities and linkages such as changes in employment, skill levels or market share as a result of innovation. Bankruptcy is a possible outcome of innovation activities and linkages which occurs when a firm fails to survive in the market. This is not always the case, as large financial institutions in the 2008–09 recession have been saved from failure by governments concerned about the consequences of bankruptcy for the stability of the financial system and employment, not just in financial services but in the entire economy.

Impacts may take time to emerge. Food services innovations leading to the rapid provision of standard food products to consumers, containing fats and sugars to make them more palatable, have been correlated with obesity, the rise of Type II diabetes and heart disease, resulting in increased demand on healthcare systems and public funds. These impacts were decades in the making. In some cases, the impacts emerge rapidly. Financial services innovations, including debt-based instruments, were introduced to the market in 2006, diffused rapidly and then lost value, requiring massive public sector intervention and the loss of income for many people. The collapse, and the economic and social impacts, happened in months.

On the positive side, the spread of broadband communications, combined with portable electronics and powerful software, have resulted in whole new industries providing web-based content to new consumers. Students go to the web before they go to the library, and libraries are changing the way they work. These changes have happened over a period of years. As will be clear from the next chapter, each example is a classic case of innovation with new goods or services, or a combination of both, being introduced to the market. The summary observation is that innovation can have both negative and positive impacts and it does not always lead to economic growth.

The use of a systems approach to understanding innovation is rooted in a broader history of attempts to understand the dynamics of systems that affect people. This work is considered briefly to lay the basis for a discussion of systems modelling in Chapter 10. The connection between systems analysis and innovation indicators is found in Gault (2007a).

System Dynamics and the Big Picture

Jay Forrester (1971, 1982) and Herbert Simon (1996) were pioneers in understanding system dynamics and it was in the System Dynamics Group of the Sloan School of Management at MIT that the original work was done that led to the Limits to Growth (LTG) project and the book of that name (Meadows et al. 1972). The project had been commissioned by the

Club of Rome and funded by the Volkswagen Foundation, and Forrester designed the prototype of the computer model that was used and contributed to the work.

The issue addressed in LTG was the sustainability of human activity, and the approach used estimates of physical limits of the carrying capacity of the planet such as the depletion of natural resources and the ability to absorb emissions from human activity. In a second book, *Beyond the Limits* (Meadows et al. 1992), the principal finding was that humanity had overshot the limits of the earth's support capacity.

This body of work and the debates it provoked, and continues to provoke was reviewed in *Limits to Growth: The 30-Year Update* (Meadows et al. 2004) and the point of raising it here is that the Limits to Growth project, while not without its critics, stimulated thought about dynamic modelling and about the data needed for dynamic modelling and the need to use systems theory in analysing complex economic and social systems.

A second reason to go back to this work now is that it exemplifies the need to address the global challenges as part of managing sustainable growth. Without addressing the big picture, which is global, complex, dynamic and non-linear in its response to policy interventions, inequities could increase and sustainability could be threatened. This point will recur in Part III in the discussion of innovation strategies.

Physical Constraints, Models and Scenarios

The debates of the 1970s around how to do dynamic modelling of economic and social systems resulted in more interest in the microdata needed to populate the variables in the models, and the addition of physical constraints took the community of scholars and practitioners beyond the natural domain of econometricians. An example was the integrated materials/energy balance statistical system (MEBSS) developed by R.U. Ayres (1978) under the auspices of the UN Statistical Office.

The MEBSS required that all material and energy inputs to the world economic system, as well as to individual countries, be accounted for either as final outputs or as changes in accumulated stocks, including durable goods in service, as well as inventories. It required two balance principles: a gross (volume) balance applied to production, consumption and trade of major resources and commodities; and a more refined materials and energy balance by a process to elucidate the relationship between production, consumption and the generation of waste flows (Gault et al. 1985).

The models of Ayres required an understanding of industrial transformation processes, a first step towards which was the need for data on the transformation of energy, material and labour into products and waste. This kind of information was not available in statistical offices although it did exist in ad hoc engineering studies. In the 1980s, Statistics Canada launched the Process Encyclopaedia Project to gather such information in a standard manner in order to support physical modelling of the economic activity (Gault et al. 1985). While some work was done, the cost of data collection and the limitations of the computing power then available rendered the project impracticable. It was abandoned in 1984.

The experience gained from the Process Encyclopaedia Project was applied to another project which supported the design of future scenarios for a socio-economic system which incorporated physical constraints (Gault et al. 1987). This was the 'design approach' to socio-economic modelling. 'Design' referred to two types of design: the engineering design that was part of the system, and the user's ability to design futures by controlling a set of variables and making choices. A fully implemented model was intended to engage the user (policy-maker) intent on setting a target of, for example, doubling the workforce engaged in R&D in five years, by making clear that there were insufficient candidates in the educational system, that it took at least ten years to produce a researcher after the start of an undergraduate programme, and that additional researchers would have to be found by other means, such as immigration. These observations were intended to start a discussion about increased immigration, conversion of workers from non-research activities to doing research, or adjusting the target or the time to reach the target to fit with the physical constraints. As with the Process Encyclopaedia, the prototype model, the Socio-Economic Resource Framework (SERF) was a learning experience. The knowledge was used to create a spin-off firm in 1989 to provide modelling services.[1]

STATISTICAL DATA

In Part II, the definitions needed to collect and interpret data on the activity of innovation are developed, followed by a discussion of the survey instruments used to collect the data and the uses of the resulting statistics. Here, a framework is presented which has been used at Statistics Canada since 1998 to guide the collection of data on science and technology (Statistics Canada 1999). However, as this is a book about innovation, the framework is adjusted to take this into account.

In 1996, the Canadian policy department, Industry Canada, funded the Information System for Science and Technology Project at the statistical office, Statistics Canada, in response to recommendations that came out of a Federal Review of Science and Technology (1994–96) (Industry Canada 1996). The purpose of the project was to produce useful indicators and a framework to tie them together into a coherent picture of science and technology activities in Canada. The resulting framework (Statistics Canada 1999) was, and is, an operational instrument for the development of statistical information on the evolution of science and technology (including innovation) and its interactions with the society, the economy and the political system of which it is a part. The framework provides a classification for science and technology activities, linkages and the related outcomes, and it makes explicit the description of the generation, the transmission and the use of scientific and technical knowledge.

The structure of the framework is given by the systems approach discussed earlier of actors, activities, linkages, outcomes and impacts, but to this is added consideration of the knowledge flows in the system, the creation, transmission and use of knowledge, including the capacity to engage in the activities of creation, transmission and use. At the time of writing in 2009, there is even more of a preoccupation with the flow of knowledge from research organizations to industry as part of commercialization; from North to South, and back, as part of development; and with how the knowledge is managed and protected.

Added to the systems approach, and the need to include knowledge, was a series of questions that supported the formulation of statements, following the discipline of the framework. The statements could also be put as hypotheses which could be tested by analysing the data being collected. The questions are basic but they encourage analytical thought and each of the variables can be given a time-dependence to support dynamic analysis.

- Activities
  - Who? Who are the actors in the system?
  - What? What is the nature of the activity engaged in by the actors, including the cost to the actor? What knowledge is produced? What knowledge is used?
  - Where? Where is the activity happening? (In a region, an industrial sector, an institution.)
  - Why? Why is the actor doing the activity? What are the objectives?
- Linkages
  - How much? What resources have been committed to the activity which involve other actors? These include expenditure, materials, energy, human resources, capital investment and knowledge transmission.

- – How connected? What are the social organizations, the support-
    ing infrastructures, discipline networks and constraints on the
    linkages of the actors?
- Outcomes
  - – What result? What happened in the short term? (Change in
      market share, skill levels of the workforce, patents, publica-
      tions, new products or processes.)

An application of the framework is the following sentence: For an actor
(Who) to perform and activity (What) in a location (Where) in order to
achieve an objective (Why), what are the costs (How much), the linkages and incentives (How connected) in order to produce an outcome
(What result)? A hypothesis could be the following: The increasing sales
of biotech firms (What result, over time) in the pharmaceutical sector
(Where), in order to gain market share (Why), is correlated with the
size of government contracts received (How much and How connected)
and the amount of refundable tax credits received (How much and How
connected).

More examples are found in Statistics Canada (1999) where the point
is made that this approach is just a beginning. In a full implementation,
a controlled vocabulary and grammar could be built up with use and
the structure elaborated, following examples from other data compilation activities (Gault et al. 1979). The quantifying of linkage information should be born in mind when discussing questions on the sources of
information for innovation discussed in Part III. As the questions are now
posed, there is no way of distinguishing between a client providing a producer of a product with a prototype for an improved product, or simply
saying that the product does not meet a list of user needs. This is a serious
consideration when it comes to studying user innovation, as opposed to
user-driven innovation.

Impacts, in the Statistics Canada framework, are dealt with by example,
as impacts appear over time and are seen in changes in economic and social
behaviour. The 2008–2009 financial crisis is one example of the impact of
innovation in financial services on the lives of people. Change in social and
business behaviour because of the use of mobile telephones is another.

The framework, developed in 1998, was rooted in the earlier systems
work of Forrester and Simon, and the knowledge component was adopted
from work of David (1993) and David and Foray (1995), and the work
leading to Foray (2004). The controlled vocabulary and implicit need
for a grammar for proposing hypotheses came from the compilation of
complex physics data (Gault et al. 1979). There was also work going on
in the same period on microeconomic simulation and dynamic analysis

(Carlsson 1997), but this was never applied in the area of innovation systems research at Statistics Canada.

In parallel with this work on capturing information about innovation and other science and technology activities in a disciplined manner, the field of innovation system analysis was evolving.

INNOVATION SYSTEMS

The concept of a 'national system of innovation' has been used as an analytical tool since the late 1980s. It has various definitions. Lundvall (1992) has argued that the definition of a national system of innovation depends on the theoretical approach. It can be narrow or broad, and he favours a broad approach which includes 'all parts and aspects of the economic structure and the institutional set-up affecting learning as well as searching and exploring – the production system, the marketing system and the system of finance present themselves as sub-systems in which learning takes place' (Lundvall 1992: 12). This broad approach fits well with the discussion of innovation policies in Part III, which is more inclusive than exclusive, and it introduces learning which has not been as prominent in innovation policy as its importance might suggest. Lundvall reviewed the work of List (1841/1959, 1909), Freeman (1987), Nelson (1987, 1988, 1993) and Porter (1990) in order to provide context to his work and that of his colleagues. The collaborators were then able to move towards a theory of innovation and interactive learning (Lundvall 1992).

The systems approach in this chapter has emphasized agents or actors, engaging in activities with other actors. This fits with the approach of Nelson, but needs some adjustment to represent the work of Freeman who looks at what the actors in the system do, and emphasizes the function (education) rather than the actor (a university). However, there is no fundamental incompatibility in the two views.

The innovation systems approach can be applied at the national, regional or local level, ranging from national systems of innovation[2] to 'clusters'. With globalization, there is more discussion of global systems of innovation. Edquist (1997, 2004) reviews systems of innovation and there is related material in Fagerberg et al. (2004). An earlier review is provided by Dodgson and Rothwell (1994) and the cluster literature is found in CEC (2008a, 2008b) along with links to the EU Innovation Strategy (Chapter 6).

Work has been done on organizational learning (Antal et al. 2001), as opposed to individual learning, and it is reviewed in Dierkes et al. (2001a), supported by an annotated bibliography of organizational learning and knowledge creation (Dierkes et al. 2001b). The creation of knowledge and its transmission and use within organizations, leading to organizational learning (Dierkes 2001), is an important part of the innovation process and a key part of innovation systems.

Cities and regions are also organizations which can learn and are part of systems of innovation. There is an extensive literature and an example is the Organization for Economic Co-operation and Development (OECD) study of Jena as a learning city (OECD 1999a). Learning regions are addressed in OECD (2002a), Florida (1998) and Wolfe (1998).

The importance of learning and the modes of learning are recurring themes in this text. The Lundvall DUI mode, learning by doing, using and interacting (Lundvall 2007), describes the activities of firms that innovate without doing R&D. It describes much of the work of user innovators, consumers or firms that are trying to solve a problem with the knowledge and tools at hand, and it also describes the interaction between users and producers in user-driven innovation. Lundvall's STI (science, technology and innovation) mode consists of science-based research processes, and fits well with larger firms able to support an R&D unit that provides new knowledge and the capacity to absorb knowledge from outside the firm.

The Lundvall and Johnson (1994) classification of knowledge into four categories is useful for understanding knowledge and learning in innovation systems. The four categories (with examples and where the knowledge might be acquired) are: know what (the density of lead: school); know why (the laws of physics: university); know how (learning by doing: workplace at start of career); and know who (learning by networking: workplace at more senior levels).

A SYSTEMS APPROACH TO A SCIENCE OF INNOVATION POLICY?

So far, a systems approach has been used to classify actors and their activities as a means of clarifying what should be measured and analysed in order to address research questions about innovation, within spatial dimensions such as cities, clusters, regions, the national and global levels, and constrained by a set of framework conditions. However, there has been limited discussion of the time dimension.

If the understanding of innovation is to improve, dynamic systems analysis is needed, which requires more investment in data and in modelling techniques. A dynamic systems model, with the linkages which provide the positive and negative feedback loops embedded in the analysis, could address some of the non-linearities of the system in its response to policy intervention or to economic shocks. This is another way of understanding the system misalignment that may result in more than one policy intervention yielding counter-intuitive outcomes. Von Tunzelman has raised various kinds of system misalignment which are important considerations in understanding how innovation policy works (von Tunzelmann 2004).

Following the proposal of Marburger, the US National Science Foundation has initiated work on the Science of Science and Innovation Policy (SciSIP). After three rounds of grants, there is one project looking explicitly at system dynamics as part of developing a science of innovation policy (Farmer et al. 2007). The separation of innovation from science here is deliberate, as innovation and science are quite different subjects calling for different policy consideration. With the problems and global challenges facing the world in 2009, a better understanding, or a science, of innovation policy is a fundamental goal, that will be reconsidered in Chapter 10.

THE SYSTEM OF NATIONAL ACCOUNTS

One of the outcomes of the depression of the 1930s was the gradual development of the System of National Accounts (SNA) as an attempt to understand the economy, its inputs and its outputs. The Second World War, and the need to manage production, provided an incentive for Leontief to develop and use input–output tables. As the SNA evolved, the knowledge was codified (CEC et al. 1994) and a language emerged that facilitated communication. The language has diffused widely and become part of common discourse as well as technical discourse. People in coffee shops are only too happy to comment on the state of gross domestic product (GDP), industrial production, merchandise trade and the balance of payments.

The innovation discussion is following the same path, but is some decades behind the SNA, and there are few coffee shops in which people can be found in animated discussion of the four components of the definition of innovation as given in the Oslo Manual.[3] The SNA is nonetheless relevant to a better understanding of innovation for at least three reasons.

Business surveys provide the information needed by the SNA on production, operating expenditures, balance sheet information, labour force, inventories and capital investment. Innovation surveys are also business surveys and benefit from the body of knowledge built up over the years by survey methodologists and by survey statisticians. These same business surveys, and related administrative data, also mean that not all information about the firm need be collected from an innovation survey if the innovation data file can be linked to other business data files in the statistical office.

SNA variables are prominent in innovation discussions. The immediate example is the division of the gross domestic expenditure on research and development (GERD) by the GDP to arrive at the most common indicator for tracking a country's progress over time, and for making international comparisons. It could be argued that R&D statistics, which are better established than innovation statistics, have been too successful in producing a single summary indicator, GERD, which can be divided by GDP to yield the GERD/GDP ratio which encourages governments to set targets, of which the Lisbon target of 3 per cent (with 2 per cent from business enterprise) is but one example. There is no single summary indicator of innovation and, given the complexity of the subject, there should never be one.

Another entry point for the SNA into the innovation discussion is through multifactor productivity (MFP), a subject of much work at the OECD over the years (OECD 2001a, 2008b). It has been argued that changes in MFP can be seen as signal of innovation (CCA 2009a, 2009b), a subject considered in Chapter 5.

A third reason why the SNA is important to the innovation discussion is that it is beginning to take account of intangibles, such as R&D, which are part of innovation. Until the UN Statistical Commission decided in 2007 that expenditure on R&D would be regarded as a capital expenditure in national accounting, it was treated simply as a business expense, not as an investment. Businesses knew that expenditures on R&D, training of staff market development were investments that would yield results over a number of years, just like a machine or a building; but this has come only recently to national accounting. The capitalization of R&D is part of a revision of the 1993 SNA which will be part the 2008 SNA. A result of the interest in intangibles is the appearance of a new OECD handbook on the subject (OECD 2009a).[4]

The capitalization of R&D has also given rise to satellite accounts linked to the SNA in a number of countries in order to understand the impact of this decision on the GDP and other economic variables. The US Bureau of Economic Analysis (BEA) reported in BEA (2007) that GDP 'would have been an average of 2.9 per cent higher between 1959 and 2004 if research and development spending was treated as investment in the US national income and product accounts'. A preliminary UK estimate suggested an increase of UK GDP by 1.5 per cent but with little impact on GDP growth (Galindo-Rueda 2007). In Canada, the distinction is made between the impact of the additional R&D capitalization, 1.6 per cent of GDP, taking account of the fact that some R&D spending is already capitalized in economic accounts (software), and the total impact of 2.9 per cent of GDP (Statistics Canada 2008a).

While much has been made of the capitalizing of intangibles in the 2008 SNA, there has been work going on for decades on the place of intangibles in the function of the firm. The work of Karl Eric Sveiby (1997) provides an entry point to this literature and it also links to the learning approach of Lundvall and his colleagues. Intangibles matter to innovation, whether or not they are capitalized in the SNA. However, as countries manage to implement the 2008 SNA, with R&D capitalized, there will be discussion of what else a firm does that could be a candidate for capitalization.

The revision of the SNA makes the point that classification systems and guidelines for the collection and interpretation of data do evolve over time, and the work of the innovation community is part of a bigger picture in which statistics are developed to advance the understanding of economic and social change. The capitalizing of R&D was an ongoing discussion in the SNA community, having been considered and rejected in the 1993 revision. In the debates leading to the 2008 revision the UN City Group, the Canberra II Group, consulted the OECD Working Party of National Experts on Science and Technology Indicators (NESTI) as part of the process, and there was strong involvement of experts from Israel. The two expert groups held joint meetings to review the problems and to move towards a productive outcome. As a result of the capitalization of R&D, the NESTI community and the SNA community will continue to exchange views about their statistical measures, the evolution of the SNA and the place of intangibles in innovation.

SUMMARY

This chapter has covered systems, scenario analysis, data requirements, national and regional systems of innovation, innovation clusters, dynamics and the science of innovation policy at a high level. It has also addressed some language issues, including a decision not to use the term 'eco-innovation systems' as a synonym for innovation systems. The objective has been to provide an overview and a context for the chapters to come, as well as references to the literature.

The 2008 SNA was introduced as a macro, rather than micro, system which contributes to the thinking about innovation and its impacts and supports macro measures of the activity of innovation in the economy. The work on the SNA also illustrated the evolution of statistical concepts and definitions and the cooperation between the SNA and the science, technology and innovation indicator communities.

Finally, the point was made that there are more intangibles in the innovation process than R&D, and that the discussion on the capitalization of intangibles in national accounts is far from over.

NOTES

1. The firm still exists: www.whatiftechnologies.com.
2. Occasionally, innovation systems are referred to as ecosystems and the rationale is that the term 'ecosystem' evokes the natural world and emphasizes the evolutionary and dynamic nature of innovation (Wessner 2007: 6). As more appears on ecological systems and green innovation, the term becomes more confusing than useful and is not used in this text.
3. As noted in the Preface, such discussion can be found in Le Passy in Paris, on occasion.
4. An overview of the issues in the capitalization of R&D and of the development of the OECD handbook is found in CES (2008a, 2008b).

PART II

Definitions and measurement

3.   Talking about innovation

INTRODUCTION

In Part II, Chapter 3 introduces the concepts and definitions needed to talk about innovation, the measurement of the activity of innovation, and the interpretation of the results. Chapter 4 introduces innovation surveys, and related surveys and case studies, that can identify innovation activity and Chapter 5 provides a discussion of the findings from innovation and related surveys. By the end of Part II, the reader should have an appreciation of what innovation is, how the activity can be measured using various means, and how the results can be used as part of the policy process and in support of institutional learning.

Experts at the OECD have been discussing innovation, its place in policy, and the need to measure it and its impacts since the mid-1980s. In the 1990s, experts in the working groups of Eurostat, the Statistical Office of the European Communities, joined in the discussion as part of managing the EU Community Innovation Survey (CIS). While the policy imperatives change from day to day, the need to measure and understand the activity of innovation remains. Over the years of discussion, a common vocabulary and grammar have emerged which facilitates the discussion and it has been described in manuals, or codified, on three separate occasions.

Manuals are like technologies and practices, they are produced and adopted, they diffuse and they can be changed by users or the users can communicate the need for change to the producers of the manuals. Users of manuals who feel that the manual does not solve their problem can develop a new manual. In this chapter, there are examples of all three activities. The chapter is about the development of the language in those manuals and its use within the Organisation for Economic Co-operation and Development (OECD) and the European Union (EU) and more recently in developing countries.

The Need for a Language

To discuss innovation a language is needed, one that facilitates an exchange of knowledge and supports peer learning. Developing such a language is not easy. It requires committed participants, time, trust and acceptance of the group consensus. However, the improved communication within the community that will result is a significant return on the investment.

Schumpeter (1934) recognized the importance of innovation in the 1930s, but it took some years to attempt to measure it and its impacts. Much of this work went on in the 1970s and 1980s with the support of the Nordic Council and with contributions of experts from Canada (C. de Bresson), Germany (L. Scholz), the UK (J. Townsend) and the US (J. Hansen). By the end of the 1980s there was sufficient experience gained by the community of practice that it was possible to start to codify the knowledge so that it could be more widely used and built upon (OECD 1992a). The mechanism used for this was the OECD Working Party of National Experts on Science and Technology Indicators (NESTI) and the result was known as the Oslo Manual.[1]

The Role of Experts

NESTI, as a group of exerts, predates the OECD. It goes back to a first meeting of experts in 1957 that gave rise to the first edition of the OECD Frascati Manual in 1963 (OECD 2002b: 151). The Frascati Manual dealt with the collection and interpretation of data on research and development but, over the years, the expert group gave rise to the 'Frascati Family' of manuals (OECD 2002c: 16) of which the Oslo Manual, dealing with innovation, was one.

The first draft of what became the Frascati Manual was prepared by Chris Freeman which is why, on the fiftieth anniversary of the first experts meeting, the book laying out the next decade of indicator development was dedicated to him (OECD 2007a). The book presented a selection of edited papers from the OECD Blue Sky Forum II held in 2006.

As an OECD working party, the membership consists of delegates from the 30 OECD member countries and the European Commission. There are also observers, such as Israel, the Russian Federation and South Africa, and other international organizations such as the UNESCO (United Nations Educational, Scientific and Cultural Organization) Institute of Statistics, the Network on Science and Technology Indicators – (Ibero-American and Inter-American RICYT) and the New Partnership for Africa's Development Office of Science and Technology (NEPAD OST). Delegates and observers are a mix of official statisticians, responsible for the development of statistical indicators, and policy analysts, responsible for the development of policy and for its evaluation once it is implemented. The mix of users and producers ensures that any outcomes of NESTI are grounded in the worlds of statistical measurement and the application of the results.

The OECD is a consensus organization, which means that the case has to be argued until delegates are convinced or, at least, will not oppose a decision. Establishing consensus ensures peer learning, which is reinforced by OECD country peer reviews of innovation policy, managed by the OECD at the request of the countries under review. Recent examples are Norway (OECD 2008b) and South Africa (OECD 2007b). Peer learning, consensus building and peer review are characteristics that make the OECD unique as an international organization and they ensure that products of the committees and working parties are used by the countries that contributed to their creation.

THE OSLO MANUAL AND DEFINITIONS OF INNOVATION

The first Oslo Manual was prepared with support from the Nordic Fund for Industrial Development and presented to NESTI in November 1989, reviewed in 1990 and sent to the Committee for Scientific and Technological Policy (CSTP) for approval in 1991. It appeared in 1992 (OECD 1992a) and it was used to guide the first European Community Innovation Survey (CIS). The Community Innovation Surveys have been reviewed by Arundel et al. (2008b) and by Smith (2004). Those surveys and innovation surveys in other countries provided an ongoing testing of the definitions and guidelines in the first edition and demonstrated the need for revision, giving rise to the second edition. The current manual is the result of the second revision.

The short review of the progress from the first to the third edition which follows illustrates the growth and importance of the common language; the role of statistical measurement, policy needs and peer learning in developing the language; and the need to go on developing the language and expanding the community of practice. The work began with technological product and process innovation in manufacturing and expanded to include non-technological innovation, and organizational and market development innovation.

The First Edition

All definitions of innovation in the Oslo manuals require a connection to the market. This is an important point considered again in Chapter 10, when public sector innovation is proposed as an item for the agenda for ongoing work. The definitions of technological innovation in the first edition were the following:

90.[2] Technological innovations comprise new products[3] and processes and significant changes of products and processes. An innovation has been implemented if it has been introduced to the market (product innovation) or used within a production process (process innovation). Innovations therefore involve a series of scientific, technological, organizational, financial and commercial activities. (OECD 1992a: 28)

92.   *Product Innovation* can take two broad forms: – substantially new products: we call this *major product innovation*; – performance improvements to existing products: we call this *incremental product innovation.* (OECD 1992a: 29)

97.   *Process Innovation* is the adoption of new or significantly improved production methods. These methods may involve changes in equipment or production organization or both. The methods may be intended to produce new or improved products, which cannot be produced using conventional plants or production methods, or essentially to increase the production efficiency of existing products. (OECD 1992a: 29)

The following were considered as a non-exhaustive list of innovative activities: research and development; tooling up and industrial engineering; manufacturing start-up; marketing for new products; acquisition of disembodied technology; acquisition of embodied technology; and design. The point was made that not all innovative activities lead to innovation, as the definition of innovation requires a connection with the market. The presence of design in the original list is noted as there was considerable interest in 2009 in measuring it. This will be discussed further in Chapter 4.

The manual went on to discuss topics to be probed by surveys including sources of information for innovation, objectives of the firm, barriers to innovation, impacts and cost. It reviewed survey methods and classifications and observed that 'the population of innovation surveys usually consists of enterprises in manufacturing industry' (OECD 1992a: 57), but does suggest that 'it may also be useful to include parts of the service sector, particularly those working directly with manufacturers'. This is a precursor to the revision leading to the second edition of the manual which included the services sector; in fact, it included the entire market economy, leaving out only the public sector (see Chapter 10).

The first revision was also happening at a time when there was a debate about how productive the service sector was and whether its impact, such as it was, was due to manufacturing firms outsourcing some of their innovation activities, such as research and development (R&D) and industrial design. This may be an explanation of the preoccupation with service firms working directly with manufacturers.

The first CIS, CIS 1, was carried out in Europe for reference year 1992 using the Oslo Manual guidelines. This was the beginning of the interaction between official surveys and the Oslo Manuals and it brought Eurostat and the OECD closer together. The second and third editions were joint productions of the two organizations.

Novelty and Technology Use

The first edition contained topics that would change or vanish in future editions. Examples are novelty of innovation and technology use surveys.

As the first edition dealt with technological innovation, it provided a classification of novelty based on aspects of technology in the innovation. It also provided the classification that would be retained in the third edition: new to the firm, the country or the market, or the world (OECD 1992a: 41), although its implementation in the CIS has been just new to the firm or to the market.

Technology use surveys, especially in manufacturing, were appearing while the manual was being developed (Ducharme and Gault 1992) and a section of the manual was devoted to them. These surveys consisted of a list of 'advanced' technologies (Statistics Canada 1987, 1989, 1991; US Department of Commerce 1989) and respondents were invited to say whether they were using or planning to use any of the technologies in the list provided. In the Canadian surveys there were questions initially on user modification of the technologies, and later (Arundel and Sonntag 1999; Statistics Canada 2008b) on adoption of the technology by developing it in-house. These questions followed the work of von Hippel (1988) and were a first probe by official statisticians of user innovation.

The Oslo Manual took a producer perspective and presented technology use surveys as measures of the diffusion of technologies produced as products by other manufacturing firms. It would take some years before the importance of user innovation would become an important policy and research question. However, the seed was there in the first manual in paragraph 185 in the sentence: 'Questions about whether the technology was modified to improve productivity or ease of use give insight into the propensity to innovate on the factory floor.'

The Second Edition

While the first CIS focused on manufacturing, it soon became evident that understanding innovation in service industries was at least as important. The often quoted statistic in 2009 is that 70 per cent of GDP comes from services in most industrialized countries and less than 20 per cent from manufacturing. The significant statistic is that half, or more, of gross domestic product (GDP) comes from marketed services and the remaining 20 per cent or so is in the public sector; education, government and health. Innovation, to be innovation, has to connect to the market, although work is being done on public sector innovation (OECD 2006a) and is being called for on consumer innovation (von Hippel 2005; Gault and von Hippel 2009). These concerns were not an issue for the innovation measurement community in 1995 when the revision of the Oslo Manual began.

In fact, discussions on measuring innovation in services had been going on for years and there was the reference already noted of such measurement in the first edition of the Oslo Manual. However, there was not the same depth of experience to draw upon as had been built up for manufacturing. This required a widening of the community of discourse and led to the inclusion of innovation in services in the agendas of Eurostat committees and of the UN City Group working at the time on service industry statistics, the Voorburg Group (Gault and Pattinson 1994, 1995). In the revision of the Oslo Manual, innovation in services was given its own working group, co-chaired by Australia and Canada.

The second edition was an improved version of the first edition, informed by survey experience and policy debate. It continued to deal with technological innovation and confined itself to product and process innovation. However it had broader economic coverage, including construction, utilities, manufacturing and marketed services. It took advantage of new international classifications, such as the 1993 revision of the System of National Accounts (CEC et al. 1994), and it recognized the importance of a systems approach to innovation (OECD/Eurostat 1997: 15) and of learning in the transfer of knowledge for innovation (OECD/Eurostat 1997: 34). Both would have a larger role in the third edition.

While the definitions remained fundamentally the same as those in the first edition, they emphasized the technological aspect of innovation. This may have reflected a view that removing or weakening the reference to technology would admit an uncontrollable flood of non-technological innovations for which the community was not ready. Here is the summary definition, which can be compared with that used in the first edition:

130.   *Technological product and process (TPP) innovations* comprise implemented technologically new products and processes and significant technological improvements in products and processes. A TPP innovation has been *implemented* if it has been introduced to the market (product innovation) or used within a production process (process innovation). TPP innovations involve a series of scientific, technological, organizational, financial and commercial *activities. The TPP innovating firm* is one that has implemented technologically

new or significantly technologically improved products or processes during the period under review. (OECD/Eurostat 1997: 47)

The definition provides an excellent example of why survey question-naires should never take their definitions uncritically from the Oslo Manual. This should not be seen as a criticism of the sometimes arcane language used. It results from lengthy debate at the end of which the use of a word, or the position of the word, may be the only way consensus is achieved. When the questions are put into surveys the language is, or should be, tested and revised before subjecting respondents to the questions.

Reference to surveys of technology use appears in the second edition from a producer perspective as a measure of diffusion. The text is essentially unchanged from the first edition, including the reference to user modification of technologies which is present in paragraph 259. The importance of learning, of knowledge and of a systems approach to understanding innovation reflected the academic literature of the time and the outcomes of the first OECD Blue Sky meeting on new science and technology indicators in 1996 (OECD 2001b).

Following the adoption in 1997 of the second edition of the Oslo Manual, and its use in the Community Innovation Surveys, the research community worked a great deal on service industries and on innovation in services (Metcalfe and Miles 2000; Boden and Miles 2000; Gadrey and Gallouj 2002; Gallouj 2002). This was not a causal relationship. This was at a time when it was becoming clear that if marketed services accounted for over half of the economy they should be better understood, and an important aspect of this understanding was how innovation in services worked.

The OECD was also engaged in innovation in services in this period, from a productivity perspective (OECD 2001c), and from the perspective of knowledge intensity and the importance of knowledge in service industries (OECD 2006b). In fact, knowledge (Foray 2007) attracted much attention in the period before the next Oslo Manual edition.

In particular, there was work on knowledge management in the business sector and its relation to innovation. A group working on this, as part of an OECD project, developed a questionnaire (OECD 2003) which had similarities to questionnaires dealing with the use and planned use of technologies. The point to make in this chapter is that the questionnaires used in the countries participating in the project worked. That is, they demonstrated that information on the use of knowledge management practices could be collected, analysed and used to improve the understanding of firm activity. Some key findings are discussed in Chapter 5.

By 2002, Eurostat and OECD were ready to undertake the three years of work needed to produce the third edition, although it was not foreseen that it would take as long as it did and be such a challenging process. The hope had been that the new manual could be used by Eurostat to guide the CIS 4. One of the lessons learned from this process was that it was difficult, if not impossible, for a consensus-based organization, with its expert group chaired by a delegate from a member country, to work to a timetable required by a supranational organization where the expert groups are chaired and directed by the Secretariat. As in all such things, it was the good will on both sides that ensured a positive outcome. It just took time.

The Third Edition

The first thing to notice about the third edition is the title of the manual, *Oslo Manual: Guidelines for Collecting and Interpreting Innovation Data* (OECD/ Eurostat 2005), and its comparison with the title of the second edition, *Proposed Guidelines for Collecting and Interpreting Technological Innovation Data – Oslo Manual* (OECD/Eurostat 1997). The word 'technological' has gone and, 'proposed' no longer appears in front of 'guidelines'. Both changes are important as non-technological innovation had now been admitted for the purposes of measurement and the Oslo Manual provided the guidelines for that measurement. The language had acquired new vocabulary.

The definition had been expanded:

> 146.   An *innovation* is the implementation of a new or significantly improved product (good or a service), or process, a new marketing method, or a new organization method in business practices, workplace organization or external relations.

It was still linked to the market through 'implementation':

> 150.   A common feature of an innovation is that it must have been *implemented*. A new or improved product is implemented when it is introduced on the market. New processes, marketing methods or organizational methods are implemented when they are brought into actual use in the firm's operations.

The definition of an innovative firm remained the same:

> 152.   An *innovative firm* is one that has implemented an innovation during the period under review.

The systems approach and knowledge management activities were incorporated in a new chapter on linkages which also addressed networks and network capital. Network capital[4] describes the knowledge stored in the networks which contributed to innovation. While the linkages chapter was a major step forward in providing guidance for the measurement of innovation, it could not deal with the dynamics of change, but it could situate the change in an innovation system.

The classification of novelty in the third edition had nothing to do with technology but was new to the firm, to the market, or to the world (OECD/Eurostat 2005: 57). There was a reference to disruptive innovation as developed by Christensen (1997), but also recognition that it was an impact measure that cannot be measured easily by an innovation survey. Disruptive innovation was not a category used for classification in the manual (OECD/Eurostat 2005: 17).

Diffusion of innovation was treated in the chapter on linkages and questions were suggested on the developer of the innovation. Was it developed by: the firm; the firm in cooperation with other firms or institutions; or mainly by other firms or institutions? This is a very important question when it comes to user innovation and it can be found in CIS 4, the CIS 2006 and in the Canadian 2005 innovation survey.

User innovation
The third edition made no reference to surveys of technology use and planned use and no explicit reference to user innovation. From the perspective of innovation surveys directed at firms, user innovation is a subset of process innovation, and the same applies to marketing and organization innovation. User innovation, in firms, is the result of the firm solving its own problems and creating new knowledge. In the case of capital equipment, this can take three forms: development of the technology; modification of an existing technology; or, just purchase of the technology. The first and second categories are user innovation. The third can be innovation if the technology purchased is new to the firm. These questions are treated in greater depth in de Jong and von Hippel (2009) and in Gault and von Hippel (2009).

The only place for the individual consumer, or end user, in the third edition is as a source of information for the firm that engages in product innovation. This is user-driven innovation as discussed in Chapter 1. The Oslo Manual does not deal with user innovation for products. This is an evolving discussion which reappears in subsequent chapters.

USING THE OSLO MANUAL IN DEVELOPING COUNTRIES

Innovation is not the prerogative of developed countries. It happens in the developing world and it can be a driver of economic growth there as elsewhere. While it may be more incremental than radical, and make more use of knowledge from sources other than R&D, it is still innovation.

Discussions took place in Latin America and in Africa about how best to measure innovation and how to produce guidelines to support the process. In Latin America, RICYT developed and published the Bogotá Manual (RICYT/OEC/CYTED 2001) and in Africa there were discussions about how to approach the need for guidelines for measuring innovation (NEPAD 2006a).

Experience with the Bogotá Manual gave rise to a proposal to the OECD to add an annex to the third edition of the Oslo Manual to interpret it for use in developing countries. This was accepted and the preparation of Annex A of OECD/Eurostat (2005) was coordinated by the UNESCO Institute of Statistics. The advantage of adding the annex to the Oslo Manual was that it could be revised, along with the rest of the manual, as experience was gained in developing countries of using both the manual and the annex. This ensured an ongoing dialogue within a broader community of practice.

In Africa, the first meeting of the African Intergovernmental Committee on Science, Technology and Innovation Indicators in Maputo in 2007 adopted the Oslo and the Frascati Manuals for use in surveying innovation and R&D activities (NEPAD 2007) in Africa. The idea was that over time, as experience was gained, African manuals could be developed to support the use of OECD manuals in African contexts (Ellis 2008; Gault 2008b; Kahn 2008).

Innovation in developing countries will be discussed in more detail in Chapter 9. The point to be made here is that African countries through the work of NEPAD OST, and Latin American countries through the work of RICYT, are using the Oslo Manual to provide guidance in measuring innovation.

MACRO SIGNALS OF INNOVATION

As discussed in Chapter 2, innovation in a firm is not an isolated event. It reflects the history of the firm, the quality of the labour force, the economic and social infrastructure of regulation, incentives, education, healthcare, telecommunications, roads, ports and culture. Innovation surveys measure the activity of innovation in a firm and the resulting data can be analysed at the firm level (OECD 2009b) or aggregated to produce population estimates for a set of indicators such as the propensity to innovate.

The System of National Accounts provided macro indicators, such as GDP, employment and trade which reflect the state of the economy and from which economic growth can be deduced. The information produced by the SNA has been used to make comparisons between Canada and the United States (CCA 2009a, 2009b). Weak labour productivity of Canada is attributed to lagging multifactor productivity which can be seen in SNA data. The use of changes in Multifactor Productivity (MFP) to detect the signal of innovation in the economy is discussed in Chapter 4. This analysis has been done with analysis using data collected under the rules of the 1993 SNA, before the capitalization of R&D (Chapter 2). It is provided as another way of detecting innovation which uses macro measures.

Innovation activities can include capital investment, expenditure on software and R&D. The first two are capitalized in the 1993 SNA. R&D is capitalized in the 2008 SNA with implications for growth measures and how the activity of innovation is seen in macro indicators. There are other intangibles that contribute to capital investment and the OECD has prepared a *Handbook on Deriving Capital Measures of Intellectual Property Products* (OECD 2009a) to provide guidance on these matters. From the perspective of this chapter, the handbook and the revised System of National Accounts provide additional language to that codified in the Oslo Manual that has to be used if all of the components of innovation are to be discussed.

SUMMARY

Since they were first discussed at the OECD in the mid-1980s, considerable advances have been made in the definition, measurement and interpretation of data on the activity of innovation. Progress has also been made on the development of international comparisons of the resulting indicators, but there is more work to be done on getting the policy community to make use of the new indicators. The community of practice has grown from OECD countries to include all EU countries, and now countries from the developing world. However, as is evident from the literature on innovation, not all researchers and fewer policy people make use of the Oslo Manual and the codified knowledge that it contains.

There are still challenges for measurement, analysis and comparison which will be discussed in Chapter 5 and again in 10. In brief, given the rapid economic changes in 2008–09, there has to be more attention paid to understanding the linkages in the system, and its dynamics. This implies the need to produce the relevant data to support firm-level microdata analysis. For analysis of system dynamics, the relevant data will come from longitudinal databases populated by survey and administrative data. As discussed later in the text, longitudinal survey data is very costly to develop and that raises questions about how scarce resources should be allocated to gain the best understanding of the innovation system.

The use of macro indicators to detect innovation is also an evolving subject as work has to be done on the development of multifactor productivity data that supports international comparisons and, even when this is achieved, thought has to be given to extracting the innovation signal from other factors that can give rise to increases in multifactor productivity.

User, as opposed to user-driven, innovation is an important part of the innovation process, both firm-based user innovation and consumer-based. User innovation and user-driven innovation have to be addressed in future editions of the Oslo Manual.

NOTES

1. The 'Frascati Family' of manuals began with the Frascati Manual for R&D statistics which was named after the town near Rome where the meeting was held that approved the manual. The name of the Oslo Manual recognized the strong support from Nordic countries for the development of the manual, and the role of Norway. There are other examples.
2. In all quotations from the Oslo Manuals, the paragraph number is included. The page number is given in the citation.
3. Oslo Manuals use vocabulary taken from the System of National Accounts (CEC et al. 1994). 'Product' refers to a good or a service. The phrase 'products and services' should never be seen in Oslo Manual-based discourse.
4. In paragraph 260 of the third edition of the Oslo Manual the observation is made that 'building social capital may be a vital part of an enterprise's innovation strategies' and then goes on to observe that 'The term *social capital* has many meanings outside of economic analysis and this can lead to confusion. *Network capital* has been used as an alternative.'

4. What can be measured?

INTRODUCTION

Since the first edition of the Oslo Manual in 1992, surveys and official statistics concerning innovation and its outcomes have evolved. The Community Innovation Survey (CIS) in Europe now covers all 27 member countries and is used in others. There have been innovation surveys in most Organisation for Economic Co-operation and Development (OECD) countries outside of the European Union (EU), in China and Russia, in African and Latin American countries and, in 2009, one began in the United States.

As a result, there has been an accumulation of data on the activity of innovation in firms, on the linkage of firms with other firms, and with other actors in the innovation system, including data on the outcomes of the activity of innovation. International comparisons are becoming established (Parven 2007; Pro Inno Europe 2009a; OECD 2007b, 2008d, 2008e). Policy use has been made of indicators derived from innovation statistics, but this is not widespread, and raises a question about the place of innovation indicators in the policy process which is discussed in Chapter 5. This chapter looks at what is being measured now and at what could be measured with the existing tools.

First, there is consideration of the measurement of innovation in 'innovation surveys', exemplified by the Community Innovation Survey for reference year 2006, CIS 2006.

This is chosen because in 2009 it is the most recent for which data are available. The questionnaire is the same as that for CIS 4 for reference year 2004, which is well documented, and the most recent European Innovation Scoreboard (EIS) 2008 (Pro Inno Europe 2009a) makes use of the CIS 2006 data. In contrast with the well-established CIS, there is the new US 2008 Business R&D and Innovation Survey (BRDIS). It is both an R&D survey and an innovation survey, having resulted from a revision of the US R&D survey, and this raises a question on the difference of measuring research and development (R&D) activities in innovation surveys and in dedicated R&D surveys addressed later in the chapter.

Second, there is a discussion of technology use surveys that uses as an example the Canadian Advanced Technology Survey (Statistics Canada 2008b) and the follow-up survey (Statistics Canada 2008c). This opens a discussion on the importance of user innovation, leading to the policy questions around what becomes of the intellectual property generated by user innovation discussed further in Chapter 5. Innovation surveys and surveys of the use of practices and technologies are complementary views of the activity of innovation.

Moving from the drawing of inferences from microdata analysis to the macro domain, there is discussion of the observation of the activity of innovation as a signal in multifactor productivity (MFP) analysis. This is based on work on innovation in the business sector in Canada (CCA 2009a, 2009b) and it emphasizes the interconnectedness of the innovation system and ways of looking at it not found in the Oslo Manual.

INNOVATION SURVEYS

The Community Innovation Survey (CIS)

As discussed in Chapter 2, there were innovation surveys long before there was an Oslo Manual. Once there was an Oslo Manual, terms like 'innovation' and 'innovation activities' were defined and recommendations were provided for what should be measured and how, and what the coverage was expected to be (industry, firm size, geography[1] . . .). This meant that boundaries were introduced in 1992, and survey statisticians, policy analysts and academics have been pushing at them ever since. This chapter looks at where the surveys are that have produced the most recent results, and the example chosen is the CIS 2006. It is based on the second edition of the Oslo Manual, not the third, which means that it is limited to product and process innovation.

The full generic questionnaire is available on the web at various sites (see Appendix A). Countries adapt the generic questionnaire to their specific needs, but the basic questions remain. A review of the history of the CIS from CIS 1 to CIS 4, and the uses of the resulting data, can be found in Arundel et al. (2008b). In what follows, the questions are listed, but without additional instructions that form part of the questionnaire.[2]

The first four questions deal with the firm and the activity of innovation as defined in the second edition of the Oslo Manual (OECD/Eurostat 1997):

General Information about the Enterprise

1. The enterprise

1.1 Is your enterprise part of an enterprise group?

1.2 In which geographic markets did your enterprise sell goods or services during the three years 2004 to 2006?

Information asked about the enterprise varies across countries, both in the first section and in the last on 'Basic Economic Information'. If the survey is being done by a statistical office with access to information about the enterprise from registers and other economic surveys, there is a case for keeping questions to a minimum and asking only about things not found elsewhere, such as the question on geographic markets. If linking the results of the innovation survey to other sources of information is a problem, there may be a case for asking one or two questions, such as those in question 12, which define the size of the enterprise and support analysis of the data by size category.

The Innovative Firm, Location of Innovation, and Novelty

2. Product innovation

2.1 During the three years 2004 to 2006, did your enterprise introduce:
New or significantly improved goods? (Exclude the simple resale of new goods purchased from other enterprises and changes of a solely aesthetic nature)
New or significantly improved services?

2.2 Who developed these product innovations?
Mainly your enterprise or enterprise group
Your enterprise together with other enterprises or institutions
Mainly other enterprises or institutions

2.3 Were any of your goods and service innovations during the three years 2004 to 2006
New to your market?
New to your firm?

Using the definitions above, please give the percentage of your total turnover in 2006 from:
Products introduced during 2004 to 2006 that were new to your market

Products introduced during 2004 to 2006 that were new to your firm
Products unchanged or only marginally modified during 2004 to
2006

3.   Process innovation

3.1   During the three years 2004 to 2006, did your enterprise introduce:
New or significantly improved methods of manufacturing or producing
products;
New or significantly improved logistics or distribution methods for
your inputs or products; or
New or significantly improved supporting activities for your processes,
such as maintenance systems of operations for purchasing, accounting,
or computing?

3.2   Who developed these process innovations?
Mainly your enterprise or enterprise group
Your enterprise together with other enterprises or institutions
Mainly other enterprises or institutions

These questions establish whether an enterprise is innovative or not, as
defined in the second edition of the Oslo Manual, and lead to the statistic
on the propensity to innovate by enterprises which can be broken down by
size of enterprise, by industry, or by geography, depending upon the level
in the firm at which the measurement is made. However, like all single
statistics, the propensity to innovate has the potential to mislead because
of what it does not convey.

Another statistic that comes from these questions is the novelty of the
innovation: whether it is new to the enterprise or to the market. Note
that no question is asked about it being new to the world. This is sub-
sumed in 'new to the market'. Not only is the novelty established, but
for products, there is a weighting factor in the form of the percentage of
turnover (revenue in North America) accounted for by the two categories
of novelty, and by those products that are not novel. In other words, all
products of the firm sold on the market are included. The percentage
of turnover accounted for by products of different degrees of novelty
has been a key economic indicator from the beginning of innovation
surveying.

Finally, there is the question of where the innovation is done: in the
enterprise, in collaboration; or in other enterprises or institutions. This
question invites follow-up questions about whether the collaborators
were users of the products produced and the place of user innovation,
discussed in Chapter 5, or about the role of public institutions in fostering
innovation.

Once innovative firms are identified, questions can be asked about their innovation activities:

Innovation Activities, Cost and Support

4.   Ongoing or abandoned innovation activities

4.1   Did your enterprise have any innovation activities to develop product or process innovations that were abandoned during 2004 to 2006 or still ongoing by the end of 2006?

Not all innovation activities lead to innovation in the reference period and some never do. The indicator, the percentage of enterprises with abandoned or ongoing innovation activities provides a firm characteristic that can be considered along with the propensity to innovate, or not.

5.   Innovation activities and expenditures and support

5.1   During the three years 2004 to 2006, did your enterprise engage in the following activities:
Intramural (in-house) R&D
    If yes, did your firm perform R&D during 2004 to 2006:
        Continuously?
        Occasionally?
Extramural R&D
Acquisition of machinery, equipment and software
Acquisition of other external knowledge
Training
Market introduction for innovations
Other preparations

This is an important question because it makes a clear distinction between innovation activities and the activity of innovation. The two are not the same. The innovation activities listed in question 5.1 do not necessarily lead to innovation. Taken in isolation, they are not innovation.

In principle, intramural R&D, and the expenditure, should be captured in the official R&D survey of the country and extramural R&D may be if there is a question on payments for R&D services. The additional question on whether R&D is performed continuously or occasionally is needed to distinguish the population of relatively rare continuous performers from the more common occasional ones. In a Canadian study (Schellings and Gault 2006) of R&D performers that were present in a nine-year period, four out of ten were there for at most two years and spent less than CAN$100000 on R&D. Occasional R&D performers dominated the population.

If the propensity to do R&D is measured by a positive response to the intramural R&D question, without taking account of the qualifying questions, the result will include the continuous performers, present for the three years of the reference period, and all of the occasional performers present for one or two years. This gives rise to a higher estimate of R&D performance than would result from dealing with just one year or with continuous performers. This is discussed later in the chapter.

The acquisition of machinery and equipment and of software should be available from the capital expenditure survey needed for the System of National Accounts (SNA). The acquisition of external knowledge, engagement in training and work on market introduction are not SNA categories, and have to be probed in the innovation survey. An exception is the acquisition of knowledge which, if it is measured by payments for R&D services or for patent licences, should appear in the balance-of-payments account in the SNA. If the acquisition is through the employment of a highly qualified person or through information received from a client, it will not appear in the SNA. In reality, it may be easier to ask about the other categories as well. The 2005 Canadian survey adds the category of 'Post-introduction commercialization' to the equivalent question (question 23).

From the perspective of user innovation, the acquisition of knowledge question identifies that this activity is happening, but it does no more than that. The receipt of the knowledge embodied in a working prototype developed by a user of a product produced by the firm is quite different from the knowledge acquired by offering maintenance services to users of the product.

Given the interest in industrial design as an innovation activity (Vinodrai et al. 2007), and its presence in the Oslo Manual from the beginning, it is interesting that it is left to the 'Other preparations' category from which it cannot be recovered. Both design and user innovation are subjects that could be probed further in CIS-like innovation surveys.

5.2   Estimate the amount of expenditure for each of the following four innovation activities in 2006 only:
Intramural (in-house R&D)
Acquisition of R&D (extramural R&D)
Acquisition of machinery, equipment and software
Acquisition of other external knowledge

This question does not cover the total cost of innovation, but it does give an indication of the magnitude of the expenditure by the enterprise on innovation activities. As innovation activities do not necessarily lead to innovation, the total expenditure should not be seen as an indicator of innovation, but of the resources allocated to the engagement in innovation activities.

> 5.3 During the three years 2004 to 2006, did your enterprise receive any public financial support for innovation activities from the following levels of government?
>   Local or regional authorities
>   Central governments
>   The European Union (EU)
>   If yes, did your firm participate in the EU 6th Framework Programme for Research and Technical Development (2003–2006)?

Once it is known that the enterprise engaged in innovation activities or not, and what it spent on a subset of them, there is a question on state support. Note that the question is limited to financial support and therefore excludes demonstration programmes like that of the National Research Council Industrial Research Assistance Program (NRC-IRAP) in Canada. The importance of this indicator is in how market failures are being addressed, by industry, geography and size of enterprise.

Information Sources and Collaboration

> 6. Sources of information and co-operation for innovation activities

> 6.1 During the three years 2004 to 2006, how important to your enterprise's innovation activities were each of the following information sources?
>   Internal
>   Within your enterprise or enterprise group
>   Market sources
>   Suppliers of equipment, materials, components, or software
>   Clients or customers
>   Competitors or other enterprises in your sector
>   Consultants, commercial labs, or private R&D institutes
>   Institutional sources
>   Universities or other higher education institutions
>   Government or public research institutes
>   Other sources
>   Conferences, trade fairs, exhibitions
>   Scientific journals and trade/technical publications
>   Professional and industry associations

Responses to this question show the relative importance of the sources, and their intensity. The client or customer always ranks high, suggesting a key role in innovation for the user of the product, but this is not elaborated upon. Universities and government institutes rank low, but the intensity goes up with increasing size of firm suggesting that this is an issue of absorptive capacity which is more likely to be present in a larger firm. While the intensity goes up, the rank order does not change substantially.

6.2 During the three years 2004 to 2006 did your enterprise co-operate on any of your innovation activities with other enterprises or institutions?

6.3 Please indicate the type of co-operation partner and location (The options are: your country; other Europe; United States; and, all other countries)

Type of co-operation partner
Other enterprises within your enterprise or enterprise group
Suppliers of equipment, materials, components, or software
Clients or customers
Competitors or other enterprises in your sector
Consultants, commercial labs, or private R&D institutes
Universities or other higher education institutions
Government or public research institutes

A collaborator is more than the information source identified in question 6.1, as the collaboration allows the exchange of knowledge among participants. Collaboration is also a linkage measure, as discussed in the third edition of the Oslo Manual.

6.4 Which type of co-operation partner did you find the most valuable for your enterprise's innovation activities?

7. Effects of innovation during 2004–2006

7.1 How important were each of the following effects on your product (good or service) and process innovations introduced during the three years 2004 to 2006? (There are four categories: high; medium; low; or, not relevant)
Product oriented effects
Increased range of goods or services
Entered new markets or increased market share
Improved quality of goods or services
Process oriented effects
Improved flexibility of production or service provision
Increased capacity of production or service provision
Reduced labour costs per unit output
Reduced materials and energy per unit output
Other effects
Reduced environmental impacts or improved health and safety
Met regulatory requirements

With the indicators so far from the survey, a picture of the activity of innovation begins to emerge and question 7.1 answers one of the remaining questions: 'What changed as a result of the activity of innovation?' The list does not deal with changes in the number of people employed or the skill levels of the labour force.

8.   Factors Hampering Innovation Activities

8.1   During the three years 2004 to 2006 were any of your innovation activities or projects:
   Abandoned in the concept stage
   Abandoned after the activity or project was begun
   Seriously delayed

8.2   During the three years 2004 to 2006, how important were the following factors for hampering your innovation activities or projects or influencing a decision not to innovate? (The importance categories were: high; medium; low; or, factor not experienced)
   Cost factors
   Lack of funds within your enterprise or group
   Lack of finance from sources outside your enterprise
   Innovation costs too high
   Knowledge factors
   Lack of qualified personnel
   Lack of information on technology
   Lack of information on markets
   Difficulty in finding co-operation partners for innovation
   Market factors
   Market dominated by established enterprises
   Uncertain demand for innovative goods or services
   Reasons not to innovate
   No need due to prior innovations
   No need because of no demand for innovations

Responses to question 8 will change according to economic conditions, but most surveys indicate that lack of qualified personnel is an inhibiting factor.

Intellectual Property Rights

9.1   During the three years 2004 to 2006, did your enterprise: (Yes or No)
   Apply for a patent
   Register an industrial design
   Register a trademark
   Claim a copyright

This is a limited intellectual property question, although it is the first appearance in the CIS of industrial design. There is no reference to other means of sharing intellectual property such as allowing free use of patents or using open source patents. These are found in the US survey discussed later in the chapter.

Organizational Innovation

10.1 During the three years 2004 to 2006, did your enterprise introduce: (Yes or No)
New *business practices* for organising work or procedures (i.e. supply chain management, business re-engineering, lean production, quality management, education/training systems, etc.)
New *knowledge management systems* to better use or exchange information, knowledge and skills within your enterprise or to collect and interpret information from outside your enterprise
New methods of *workplace organisation* for distributing responsibilities and decision making (i.e. first use of a new system of employee responsibilities, team work, decentralisation, integration or de-integration of departments, etc)
New methods of organising *external relations* with other firms or public institutions (i.e. first use of alliances, partnerships, outsourcing or sub-contracting, etc.)

10.2 Who developed these organisational innovations?
Mainly your enterprise or enterprise group
Both your enterprise and other enterprises or institutions (including consultants)
Mainly other enterprises or institutions (including consultants)

10.3 How important were each of the following effects on your enterprise's organisations innovations introduced during the three years 2004 to 2006? (High, medium, low or not relevant)
Reduced time to respond to customer or supplier needs
Improved quality of your goods or services
Reduced costs per unit output
Improved employee satisfaction and/or lower employee turnover
Improved communication or information sharing

Marketing Innovation

11.1 During the three years 2004 to 2006, did your enterprise introduce the following marketing innovations (Yes or No)
Significant changes to *product design* or the packaging of goods or services (exclude changes that only alter the product's functional or user characteristics)
New media or techniques for *product promotion* (i.e. the first time use of a new advertising media, fundamentally new brand to target new markets, introduction of loyalty cards, etc.)

New methods for *product placement* or sales channels (i.e. first time use of franchising or distribution licences, direct selling, exclusive retailing, new concepts for product presentation, etc.)
New methods of *pricing* goods or services (i.e. first time use of variable pricing by demand, discount systems, etc.)

11.2 Who developed these marketing innovations?
Mainly your enterprise or enterprise group
Your enterprise together with other enterprises or institutions (including consultants)
Mainly other enterprises or institutions (including consultants)

11.3 How important were each of the following effects of your enterprise's marketing innovations introduced during the three years 2004 to 2006?
Increased or maintained market share
Introduced products to new markets or customer groups
Increased visibility of products or business
Improved ability to respond to customer needs

Questions 10 and 11 anticipate the expansion of the definition of innovation to include organizational change and practices and market development. Their introduction here allowed them to be widely tested before they could be considered as part of a four-part innovation question. Questions 10.2 and 11.2 also probe for the presence of user innovation.

Basic Economic Information on Your Enterprise

12.1 What was your enterprise's total turnover for 2004 and 2006?

12.2 What was your enterprise's total number of employees in 2004 and 2006?

That completes the generic version of a CIS 2006 survey of enterprises. It is on the basis of such a survey, conducted in European countries, that all of the innovation indicators discussed in later chapters are derived. Links to the Canadian surveys are found in Appendix A.

The US Business R&D and Innovation Survey (BRDIS)[3]

After years of absence from the measurement of innovation, the US National Science Foundation (NSF), in collaboration with the Economic Directorate of the Bureau of the Census, redesigned the R&D survey to produce the Business R&D and Innovation Survey (BRDIS) which went into the field as a pilot survey on 26 January 2009. The stratified sample of 40 000 firms with five or more employees includes a census of large R&D performers; the 50 largest firms, based on payroll, in each state; and a sample of other firms drawn from the Census Business Register (US Census Bureau 2009).

The timing of this initiative anticipated the release of the US Innovation Strategy (Executive Office of the President 2009) and the NSF should be in a position to deliver policy-relevant results as the new strategy is being implemented. The BRDIS results will transform US R&D statistics as well as contributing to policy development (*Business Week* 2008). The survey will also produce official statistics on innovation in US firms based on the results of the following question:

6.1   Did your company introduce any of the following during the three-year period, 2006 to 2008?
   a.   New or significantly improved goods (excluding the simple resale of new goods purchased from others and changes of a solely aesthetic nature)
   b.   New or significantly improved services
   c.   New or significantly improved methods of manufacturing or producing goods or services
   d.   New or significantly improved logistics, delivery, or distribution methods for your inputs, goods, or services
   e.   New or significantly improved support activities for your processes, such as maintenance systems or operations for purchasing, accounting, or computing

This question is a combination of 2.1 and 3.1 used in the generic CIS 2006 questionnaire and the responses will support some comparison with CIS and CIS-like survey findings. The BRDIS is a pilot survey as far as innovation is concerned, and it is designed to be the platform for future in-depth modules on innovation in industry. The questions used in 2009 are being revised and added to for the 2010 survey so that there will be more information on the activity of innovation in the US in the coming years.

The innovation question is followed by a series of questions on intellectual property, including one on intellectual property transfer activities. In that question there are nine options, but the last two on free revealing are those of interest in the discussion of user innovation in Chapter 5. The full question follows:

6.9   Did your company perform the following activities in 2008?
   a.   Transferred intellectual property to others not owned by your company through participation in technical assistance or 'know how' agreements

b. Received intellectual property from others not owned by your company through participation in technical assistance of 'know how' agreements

c. Transferred intellectual property from a parent company as part of a spin-off or spin-out

d. Acquired more than 50% ownership in another company for the primary purpose of acquiring their intellectual property

e. Acquired any financial interest in another company in order to gain access to their intellectual property

f. Participated in cross-licensing agreements – agreements in which two or more parties grant a license to each other for the use of the subject matter claimed in one or more of the patents owned by each party

g. Allowed free use of patents or other intellectual property owned by your company (for example, allowing free use of software patents by the open source community)

h. Made use of open source patents or other freely available intellectual property not owned by your company

Measuring R&D in Innovation Surveys

The examples of the CIS and the BRDIS raise a question about the measurement of R&D in surveys. The R&D question (5.1) in the CIS confirms, over the three years covered by the survey, the presence of the performance or acquisition of R&D, and the response to the performance question supports estimates of the propensity to do R&D. Two observations follow from this. The first is that more firms innovate than do R&D on a full-time basis (Arundel et al. 2008; OECD 2009b). The second is that the estimate of the propensity to do R&D found in innovation surveys is higher than the propensity to do R&D found in dedicated R&D surveys based on the definitions in the Frascati Manual (OECD 2002b).

Country comparisons of non-R&D innovation propensities are given by Arundel et al. (2008a) for CIS 3 results and results from CIS 4 and similar surveys in non-EU countries can be found in OECD (2009b) Tables S.3 for innovation and S.13 for R&D performance. Comparing the two tables in OECD (2009b) shows that in all countries in the study, with two exceptions, the propensity to innovate exceeds the propensity to do R&D. The exceptions are Japan and Korea. The comparisons are given in Table 4.1.

A possible explanation of these results could be a large population of occasional R&D performers, as question 5.1 asks about the performance of R&D over the previous three years and whether it is continuous or occasional.[4] If there is a large population of occasional performers, with different ones appearing in each of the three years covered, this statistic would be higher than if the question was put for just one year. A better measure of R&D propensity could be a count taken from the response

Table 4.1   International comparison of the percentage of firms in manufacturing engaged in R&D and in innovation

	R&D (%)	Innovation (%)	Difference
Austria	32.7	55.4	22.7
Belgium	35.2	54.0	18.8
Canada	53.2	65.0	11.8
Denmark	27.7	51.3	23.6
Finland	37.9	44.8	6.9
France	27.7	35.0	7.3
Germany	47.3	65.9	18.6
Japan	27.9	24.4	−3.5
Korea	42.0	40.2	−1.8
Luxembourg	27.4	47.2	19.8
Netherlands	29.6	39.5	9.9
New Zealand	19.0	48.0	29.0
Norway	32.4	36.3	3.9
Sweden	40.7	51.3	10.6
Switzerland	47.9	67.2	19.3
United Kingdom	40.2	41.9	1.7

Source: OECD (2009b), Table S.3 (Firms having introduced a product or process innovation) and S.13 (Firms that perform R&D); and author calculations.

to question 5.2 on expenditure on intramural R&D performance for the single year, 2006.

To make the point about dedicated R&D surveys, the Canadian results in OECD (2009b) are for manufacturing and only for firms with 20 or more employees and a turnover of more than CA$250 000. The innovation propensity is 65 per cent, the R&D propensity from the same survey is 53 per cent (Uhrbach 2009) and the result from the same population of respondents in the R&D survey is 34 per cent (Government of Canada, Science, Technology and Innovation Council 2009: 15). As the total number of R&D performers in Canada increased by 76 per cent in the period between 2000 and 2005, a compound annual growth rate of 12 per cent, there is a need to examine the amount of R&D that new entrants are actually performing in order to understand the significance of the 34 per cent (Statistics Canada 2009, Table 19.2). However, it can be regarded as an upper bound from the R&D survey which is still significantly lower than the 53 per cent found in the innovation survey.

The US BRDIS is a revision of the US R&D survey and the expectation is that the propensity to do R&D found in the BRDIS will be significantly lower than that found in CIS and CIS-like surveys. In Chapter 5, the policy implications of innovation by non-R&D performers are discussed.

TECHNOLOGY USE SURVEYS AND INNOVATION

In the last chapter, there was a brief discussion of technology use surveys and their role in measuring innovation, and reference to them in the first two editions of the Oslo Manual. Here the subject is approached from a survey perspective. The example of a technology use survey is the Statistics Canada Advanced Technology Survey 2007 (Statistics Canada 2008b) and the follow-up survey (Statistics Canada 2008c). The questionnaires and survey methodology can be found on the Statistics Canada website (see Appendix A).

As described in OECD (1992a), technology use surveys are simple. They consist of a list of technologies or practices and the respondent is asked to say if they are used, planned to be used or not planned to be used. A technical manager is able to look around the plant and answer the questions without consulting records as the question is use, not capital expenditure. This means that the response rate for such surveys is high as, properly designed, they should be easily understood by the plant manager.

The questionnaires have a section on technologies and a section on factors related to their adoption, similar to those used in the innovation surveys just discussed: percentage of the capital expenditure budget spent on advanced technologies; skill requirements; sources of information or assistance; outcomes of the adoption; and obstacles to adoption. In the Advanced Technology Survey (Statistics Canada 2008b), question 4 asks:

> How does your business unit acquire or integrate advanced technologies (equipment and/or software)? Please check all that apply:
> - By purchasing off-the-shelf advanced technology (equipment and/or software)
> - By leasing off-the-shelf advanced technology (equipment and/or software)
> - By licensing advanced technology
> - By customizing or significantly modifying existing advanced technology
> - By developing new advanced technologies
>   (Either alone or in conjunction with others)
> - By merger or acquisition of another firm with advanced technologies.

The important questions here are whether the plants adopted technologies by acquiring (purchasing, leasing or licensing), modifying or developing technologies. The merger and acquisition question is separate from this discussion.

There is the view expressed by von Hippel (2005), implicit in the first two editions of the Oslo Manual, that process innovation is producer driven. That is, all of the high-level innovation takes place in the firm of the producer of the technology and the technology is then sold to users. The users may qualify as innovators if they make the purchase within the reference period of the innovation survey and the application of the technology is new to the firm, the lowest level of novelty that qualifies as innovation.

If innovation is producer driven, the expectation is that the intellectual property of the producer will be protected with the existing range of instruments: patents, copyrights, trademarks, and registration of design, seeds or circuits. The knowledge about the technology is expected to flow back to the producer firm through service agreements or interaction with the marketing staff, ensuring that users (customers or clients) are placed high on lists of sources of information that influence the innovation of the producer. There is more to it than this. Arundel et al. make the point that:

> an important method (not identified in the CIS surveys) that is used by both non-R&D innovators and R&D performing firms to innovate is to customize or modify products, processes, or organizational methods developed by other firms or organizations. This is reported by approximately one-third of both types of firms. (Arundel et al. 2008a: 32)

This supports the view that there is a significant presence of user innovation. Table 4.2, taken from von Hippel (2005), further illustrates the importance of the user role in innovation, both as part of the production process and as end users or consumers.

More recent results in support of user innovation come from a collaboration of Eric von Hippel and Christoph Hienerth[5] on user innovation in rodeo kayaking. Rodeo kayaking was founded by users, and has followed an innovation trajectory typical for many user-developed sports. During the course of its development, many important and novel techniques and interdependent novel kayak products were developed and sold to practitioners of the sport. Studies of the histories of all these innovations showed that 63 per cent of the major product innovations and 83 per cent of the minor product innovations functionally important to the sport were developed by users rather than producers. User-innovators were also responsible for 100 per cent of the technique innovations that utilized and induced those product innovations (Hienerth 2006).

The major finding with respect to innovation expenditures is that the collective investment in product development by rodeo kayak users is much larger than that of all rodeo kayak producers. Indeed, conservative analyses show that aggregate innovation investment by kayak users is larger than the aggregated investment in the development of kayaks and

Table 4.2   Studies of user innovation frequency

Innovation area	Number and type of users sampled	Developing and building product for own use (%)
Industrial products		
1. Printed circuit CAD software	136 user firm attendees at a PC-CAD conference	24.3
2. Pipe hanger hardware	Employees in 74 pipe hanger installation firms	36
3. Library information systems	Employees in 102 Australian libraries using computerized OPAC library information systems	26
4. Medical surgery equipment	261 surgeons working in university clinics in Germany	22
5. Apache OS server software security features	131 technically sophisticated Apache users (webmasters)	19.1
Consumer products		
6. Outdoor consumer products	153 recipients of mail order catalogues for outdoor activity products for consumers	9.8
7. 'Extreme' sporting equipment	197 members of 4 specialized sporting clubs in 4 'extreme' sports	37.8
8. Mountain biking equipment	291 mountain bikers in a geographic region known to be an 'innovation hot spot.'	19.2

Sources: Data from: (a) Urban and von Hippel (1988); (b) Herstatt and von Hippel (1992); (c) Morrison et al. (2000); (d) Lüthje (2003); (e) Franke and von Hippel (2003); (f) Lüthje (2004); (g) Franke and Shah (2003); (h) Lüthje et al. (2002).

kayak-related products by producers. In the case of technique, the investment ratio is even more lopsided. Producers made essentially no investment in technique – and technique determines the functionality delivered by sporting products.

The implications of these findings, if they can be generalized, are that analysts and policy developers need to take account of the true cost of innovation and where that cost was incurred. This is relevant to analysis based on the industry of production. An innovation survey of kayak producers would see the introduction of new products to the market and the importance of the end user as a source of information for innovation. The cost, as measured by CIS question 5.2, would be quite low as no R&D would be required to take the prototype from the consumer. Costs would only appear if the consumer charged for the knowledge that was being transferred, or if new machinery, equipment or software were required to produce the improved kayak.

When the CIS question 2.2 is asked about who developed these product innovations, the likely answer would have to be: 'Mainly other enterprises or institutions'. The problem here is that the kayak user, appearing before the producer with a significantly improved kayak, might not be perceived as an enterprise or institution. This is a point to be considered in the wording of subsequent versions of this question, and in the reporting guide that goes with the questionnaire.

From the policy perspective, there may be a case of supporting the use of university or government laboratories for kayak development if the final product was commercialized by a producer. There could also be voucher schemes, like those in the UK or the Netherlands, allowing the producer or the users to buy advice and other services as part of building a better kayak. Of course the word 'kayak' here could be replaced by any other consumer product that has engaged the imagination and the financial resources of the consumer.

Getting back to innovation in firms, in earlier surveys of technology use and planned use a significant percentage of the target population indicated that they either modified or developed advanced technologies for their own use. The 2007 Statistics Canada survey confirmed this, with 21 per cent of users of at least one of the technologies surveyed adopting by modifying a technology and 22 per cent adopting by developing a technology. This was a significant signal of user innovation as these were population estimates for the universe of manufacturing plants, not case study evidence. A classification of innovators, including user-innovators, is found in Tables 1.1 and 1.2.

The pilot follow-up survey (Statistics Canada 2008c) addressed questions about the modification or development to a subset of respondents who had revealed themselves as user-innovators by being modifiers or developers. The questions are available on the Statistics Canada website and the complete responses, with analysis, are given by Schaan and Uhrbach (2009); an analysis from a policy perspective is found in Gault and von Hippel (2009).

Surveys of advanced technologies in manufacturing are only part of the picture. Since 1998 the OECD has encouraged measurement of the use of information and communication technologies (ICTs) (OECD 2009c), biotechnologies (OECD 2009d) and nanotechnologies (OECD 2009e).

Surveys in these areas have not been focused on innovation, but on the use of new or emerging technologies as part of the production process or as products. Nonetheless, firms using these technologies would be classified as innovative in an innovation survey.

MANAGEMENT PRACTICES USE SURVEYS AND INNOVATION

Management practice is a broad topic and only a subset of practices will be considered here: those that deal with knowledge management, and only in the business sector. The key point to be made is that surveys to establish the use and planned use of knowledge management practices are no different from surveys of the use and planned use of technologies (Earl 2002, 2003). Both practices and technologies are ways of doing things.

The second point is that an OECD working group which involved the statistical offices of Canada, France, Italy, the Netherlands and Sweden with representatives from Australia, Denmark, Germany and Ireland met in Copenhagen, Ottawa, Paris and Karlsruhe and developed a questionnaire which was used in a number of the participating countries. The results, and the questionnaire, were published in OECD (2003). This exercise established that knowledge management practices could be studied in the same way as technologies, and that there were correlations between the use of such practices and the activity of innovation (Kremp and Mairesse 2002).

While practices are more flexible than technologies, they can exhibit exactly the same types of innovation as described in Chapter 1 in Tables 1.1 and 1.2. Practices can be producer-driven innovations, user-driven innovations or user innovations. The parallel between technologies and practices has been explored by Gault and McDaniel (2002), but the topic has received little attention by the innovation community.

CROSS-SECTIONAL OR PANEL SURVEYS

All of the surveys discussed in this chapter are cross-sectional surveys. That means that the data are collected for one reference period, which may be a year for technology use surveys or a three-year period for innovation surveys. The results provide a snapshot of the period and repeated snapshots provide changes in aggregate statistics over time, but they do not support inferences of causal relationships, only correlations, an example of which is that between firm size and the propensity to do R&D or to innovate.

There are also longitudinal surveys in which firms, constituting a panel, are measured repeatedly over a period of time. The data from these surveys can support inferences about causal relationships, but at considerable cost. The cost includes the burden on the participants, which have to commit to being in the panel for a period of time, and the infrastructure needed to maintain the survey and the more complex analysis. The reporting burden can be reduced by the use of administrative data, if the survey agency has access to such data. Panels suffer from the deaths of firms, from mergers and acquisitions, and from refusal to continue to make the considerable contribution required to the public good. This means that the support team has to have the means to renew the panel membership. While there are many panel studies in the social and behavioural sciences, they are not often found in the production of official statistics for innovation analysis. Where they do occur is in business conditions surveys, market assessment surveys and intentions surveys, where a small number of focused questions can be posed. An interesting exception is the Mannheim Innovation Panel (MIP) (Janz et al. 2001) which is part of the German CIS and is used as a research tool at the Centre for European Economic Research (ZEW), Mannheim. Panel surveys have been used in New Zealand (Fabling 2007) and Statistics Canada is initiating a business panel survey, the Survey of Innovation and Business Strategy. Going back to earlier discussion, it will have a question on user innovation.

MULTIFACTOR PRODUCTIVITY

Innovation surveys and surveys of the use of technologies and practices support both industry-level and firm-level analysis. At a more macro level, the System of National Accounts (SNA) brings together estimates and data from all parts of the economy and is able to calculate multifactor productivity (MFP) measures that can be used, in principle, for international comparison (OECD 2001a, 2008b). In Canada it is of interest as lagging MFP is seen to be primarily responsible for the weak trend of labour productivity (CCA 2009b: 34).

MFP is not measured directly but is inferred as a residue that measures the portion of labour productivity growth that cannot be accounted for by measuring the growth of capital intensity and the quality of the workforce. A recent Canadian report makes the point that:

> MFP growth contains the macroeconomic signature of aggregate business innovation – the extraction of increasing value from inputs of capital and labour through inventive activity, entrepreneurship, the more efficient

organization of work, new marketing practices and business models, the payoff from performing R&D and the capture of the benefits of innovation originating elsewhere and particularly the insights of entrepreneurs. (CCA 2009b: 34)

While MPF estimates are fraught with data problems, changes in the estimate over time reflect the impact of all of the policy interventions of government and of the environment in which innovation happens, or does not happen. The measure also relates to economic growth, which is one of the policy objectives of promoting innovation.

Developing standard methods for calculating MFP that would lead to results which could be compared across countries is a subject for future work (Chapter 10). There is still the challenge to find ways to disentangle signals of innovation from those arising from other economic and social activities. It is also clear that this is a macro indicator which can never address firm-level issues such as how firms innovate without doing R&D, and what gives rise to user innovation.

SUMMARY

This chapter has examined what has been measured in innovation surveys and in the complementary surveys of the use and planned use of technologies and practices. This has shown how the fact that firms are innovative, or not, is inferred from the responses to the questions about their activities. No firm is ever asked if it is innovative.

One of the more robust observations from innovation surveys is that the propensity to innovate is greater than the propensity to do research and development. This raises questions about how non-R&D performing firms manage the activity of innovation, which will be discussed further in Chapter 10.

Existing results from business surveys demonstrate that user innovators can be identified as a sub-population of innovators, and suggestions will be made in Chapter 5 about ways of gaining more information about their existence. There is little information from official statistics on the end user as innovator and how the effect of this can be measured. Some early results from Hienerth and von Hippel discussed in this chapter suggest ways of getting at user innovation of products. This information could have policy implications with far-reaching consequences.

Finally, the existence of an innovation signal in changes in multifactor productivity over time was discussed along with the data problems and the resulting implications for international comparisons of MFP data.

NOTES

1. Geographical breakdowns of survey results can present a problem, depending upon how the sample is drawn. Sampling at the firm or enterprise level supports geographical breakdown only for firms that have one location. Such firms tend to be small and medium-sized enterprises (SMEs). Larger firms are likely to have production activities in more than one location and in more than one industrial classification. In the case of the Canadian Innovation 2005 Survey the sampling took place at the establishment or plant level. The reason for this was that establishments almost always have one location and that meant that the result could be distributed geographically for the whole sample and estimates provided for the survey universe, where accuracy and data quality permitted. See Appendix A for direction to the Innovation 2005 survey methodology.
2. This distinction between survey questions and a survey questionnaire is worth noting. Questions are developed, ideally in consultation with policy experts from the department that will use the results in policy development or evaluation. Once they are developed, survey statisticians structure the survey to maximize information gathered by the questionnaire while keeping the reporting burden to a minimum. The outcome of this activity is a questionnaire which will contain what are called skip patterns (if the respondent answers no to question 7, they are sent to question 21, not to question 8). The questionnaire is then tested in the appropriate languages, and on the basis of that testing some questions may be discarded and the questionnaire redesigned. Of course, this is an ideal situation.
3. Lynda Carlson and John Jankowski provided information for this section.
4. Vladimir López-Bassols from the OECD and Francois Rimbaud and Pierre Therrien from Industry Canada were helpful in clarifying this issue.
5. Eric von Hippel and Christoph Hienerth were kind enough to share their work at a preliminary stage.

5. How are indicators used?

INTRODUCTION

Chapter 3 developed a language for the discussion of innovation and Chapter 4 provided examples of the application of that language to the development of surveys and the interpretation of their results. This chapter looks at how those results can be used.

An observation made by Arundel (2007) is that innovation indicators are not used for policy purposes even though they have been available from several rounds of the Community Innovation Survey (CIS), the first being for reference year 1992. Various explanations are provided that include the dominance of well-established research and development (R&D) incentive programmes, and the Lisbon target of 3 per cent of gross domestic product (GDP) to be allocated to R&D. In the US in 2009, there were no official statistics on innovation to use to support policy analysis, but this is changing. Over the years, academic research has provided little policy guidance, and country comparisons of innovation activities have been made difficult by problems of accessing the data.

Arundel (2007) also stresses the importance of innovation not based on R&D, and cites the Aho Report (CEC 2006a) and the Competitiveness and Innovation Framework Programme (CIP) (CEC 2005) to support the importance of the diffusion and application of technologies which may not necessarily involve R&D. The diffusion and application of technologies (and practices) raises the significance of the role of user innovation, which has not been part of the innovation policy debate. User innovation and non-R&D-based innovation are also relevant to development policy, as well as to domestic policy in developing countries, a subject which will be addressed in Chapter 9.

Indicators based on statistics populated by data can provide information about the state of a system. Some examples are the propensity to innovate in an industry or region, or the expenditure on R&D, or the number of patents filed by firms in the industry. Repeated measurements support monitoring of the behaviour of the system. At the programme or project level such measurements, when combined with a set of performance criteria, can be used for evaluation. Indicators can be used for benchmarking exercises where the present state of the system is compared with a desired future state, or with the state of another system, where the objective is to move from the initial state to the target state. Finally, indicators can be used to support foresight exercises. All of these applications are reviewed in order to illustrate the different ways of using indicators as part of the policy process. However, indicators can also be misused and misinterpreted and the chapter offers warnings about the use of indicators. Finally, indicators need to be understood in order to be used, which calls for better absorptive capacity in policy departments.

USING INDICATORS

Supporting Policy Learning

Monitoring

The most benign use of indicators is the monitoring of the innovation system by comparing the values of a set of indicators over time. The European Innovation Scoreboard (EIS) supports monitoring of European Union (EU) countries; the Global Innovation Scoreboard extends the comparison of the EU27 to high-R&D-performing countries (Pro Inno Europe 2009a). The OECD publishes the Main Science and Technology Indicators (OECD 2008e) twice yearly, and every two years the OECD Science, Technology and Industry Scoreboard (OECD 2007c) and the *OECD Science, Technology and Industry Outlook* (OECD 2008d). In addition, as the result of a microdata analysis project, a set of innovation indicators have been used for the first time to make intercountry comparisons (OECD 2009b). In the US, the Science and Engineering Indicators are published every two years (US National Science Board 2008), as is the indicator report of the Observatoire des sciences et techniques, published in France (OST 2008). Germany publishes the *Federal Report on Research and Innovation* (BMBF 2008a) every two years.

Not all of the indicators in these publications are indicators of the activity of innovation as defined in the Oslo Manual, but there are indicators of innovation activities such as R&D performance, capital investment, intellectual property protection, learning, education and design. The indicator reports can be supplemented by economic data produced by the System of National Accounts. All of these indicators contribute to the public policy debate on science and innovation.

Indicators can be used to monitor public spending on science, technology and innovation (STI) and to answer three questions:

- How much does the government spend on STI?
- Where does it spend it (geography and industry)?
- Why does it spend it (socio-economic objectives)?

A fourth question, 'What does the government get for spending this money?' requires a systems approach to get close to a meaningful response and, if coupled with the policy objectives of government, it becomes a topic for evaluation (Gault 1998).

An example of an attempt to answer the fourth question is the work on the Science of Science Policy (SoSP) being undertaken by the US government and described in *The Science of Science Policy: A Federal Research Roadmap* (NSTC 2008). The text sets out three broad themes: understanding science and innovation; investing in science and innovation; and using the SoSP to address national priorities. It is the last theme that tries to get at the fourth question. The report notes that the US data infrastructure is inadequate for decision-making and suggests that the research on SoSP could be used to make better R&D management decisions and to quantify the impact that the scientific enterprise has on innovation and competitiveness.

This is a major undertaking in indicator development, monitoring and evaluation with a view to quantifying outcomes of government investments. It could be regarded as the federal component of the National Science Foundation (NSF)-supported Science of Science and Innovation Policy (SciSIP) which involves the academic community and analysis of private sector activity.

Benchmarking

Monitoring the system is one thing: deciding where it should be going and how it is to get there is another. There are two broad approaches to benchmarking. The first is to decide upon a set of indicators which are relevant to policy objectives. Once they are agreed, targets can be set and, ideally, the move to the target is supported by policies and programmes. A second approach is to agree upon the set of indicators and then select another system that may be in other respects comparable but is performing better according to a set of performance criteria. The values of the set of indicators for the comparable system become the targets. The advantage of the second approach is that it includes a dynamic element as the comparator system may react quite differently to economic shocks, such as the recent crisis.

An example of the first approach is the Canadian Innovation Strategy, released by a former government in 2002 as two papers, one from Industry Canada (Industry Canada 2001) and one from the Department of Human

Resources Development Canada (HRDC 2002). The Industry Canada paper provides examples of targets, a selection of which follow. By 2010:

- rank among the top five countries in the world in terms of R&D performance;
- at least double the amount invested in R&D by the Government of Canada; and,
- rank among world leaders in the share of private sector sales from new innovations.

The question is not whether these were realistic targets in 2002, or not. It is an example of setting targets in order to focus public debate and government policy.

The simplified version of benchmarking is to set a single target, such as the Lisbon target of 3 per cent of GDP (2 per cent to come from the business sector) to be allocated to R&D by 2010, and then to provide policies at the EU level and encouragement at the country level to develop complementary policies and to share best practices which have been identified through case studies or analysis of relevant survey or administrative data.

Evaluation
Evaluation concerns the effective and efficient allocation of resources in order to achieve a set of objectives. To be of use it has to be done at a level, such as the project or programme, where both inputs and outputs can be measured. There are various methods of evaluation ranging from the quantitative (bibliometric analysis, turnover resulting from new products introduced to the market, audits and so on) to the qualitative (such as peer review), and there are mixtures of the two. Innovation indicators can be used as part of evaluation. An example at programme level is the evaluation of the US Small Business Innovation Research (SBIR) programme (Wessner 2008).

One form of high-level evaluation is the Country Reviews of innovation policy conducted by the OECD. The reports are a mix of case studies, interviews and analysis following a practice built up over years of experience. On the basis of the evaluations, countries may revise their innovation policy mix. UNU-MERIT (United Nations University Maastricht Economic and Social Research and Training Centre on Innovation and Technology) offers a course on Design and Evaluation of Innovation Policy in Developing Countries (DEIP) which acquaints participants with the components of innovation policies, how they are used, and how the results can be monitored and evaluated.

Innovation strategies, as with any policy initiative, should have monitoring and evaluation built in to the process. This is necessary to assess progress and signal the need to change direction in response to what is found, or in response to economic shocks.

Foresight

Foresight is an exercise in viewing the future and there is no simple definition of the activity, a point made by Miles et al. (2008: 3). As with evaluation, foresight can involve a mix of quantitative and qualitative methods, including the use of a current set of indictors. Georghiou et al. (2008) discuss the history of foresight and its evolution and the methods employed.

In the UK, the Technology Strategy Board has developed a series of technology strategy papers, and in each case the technology is linked to the demand side. This is seen to be a key element in future government action. From the foresight perspective, action plans follow the strategies developed by working with business to build a 'road map' to plan what government and business will do together in the future.

The OECD has also engaged in foresight, or futures studies, and an example that touches upon the topics in this book is *The World in 2020* published in 1997 (OECD 1997). Other examples that make interesting reading in 2009 are *21st Century Technologies* (OECD 1998) and *The Future of the Global Economy* (OECD 1999b).

Foresight can be used to discern likely paths for emerging technologies, possible futures for an economic region, or alternative responses to a coming problem such as how the country should function when the oil runs out. Indicators in general, and innovation indicators in particular, provide a background to such a process, but for most problems addressed by foresight the indicators are too aggregated to be a major part of it. These are higher-level issues than those addressed in this text, which looks at how the indicators of innovation are developed and used in the policy process. The focus is on the present, informed by the past, perhaps with targets for the future. Foresight, with its emphasis on what the future is, and on how to get there, is outside the scope of the text.

Supporting Learning about Business Behaviour

Many of the characteristics of firms that support learning about business behaviour were covered in Chapter 1 in the section on stylized facts. Only one example is given here as it is not that well known, but it does have implications for understanding non-technological innovation. The example is the use of knowledge management processes, and there are two findings: their use is correlated with the presence of the activity of innovation; and the variables describing the practices exhibit discontinuities as the size of the firm changes.

As part of the OECD project on knowledge management in 2001, and the design of a survey of the use of the practices (OECD 2003), some questions were inserted into the CIS 3 in France and a correlation was found between the use of knowledge management practices and innovation (Kremp and Mairesse 2002). This led to discussions on the importance of non-technological innovation in general, and of knowledge management in particular (de la Mothe and Foray 2001).

In Canada, there was a pilot survey of the use of knowledge management practices which showed that those practices that suited a small firm did not continue as the firm grew, but diminished in use while practices more appropriate to a larger firm took over. As an example, the coffee pot ceased to be the centre of knowledge sharing in a firm of ten people, to be replaced by management meetings and the circulation of newsletters and reports for firms of 250 people. What was interesting was the size at which the transition took place (Earl and Gault 2003).

Making a Case

One of the uses of indicators is to make a case for further analysis leading to policy development. Two examples are provided: user innovation and the career paths of doctorate holders.

User innovation

The first task is to establish the magnitude of the phenomenon. User innovation is problem-solving by the user, or would-be user, involving technologies or practices, that results in better or new technologies or practices. User innovation is not measured directly in CIS-type surveys, although there is an indication of its presence which will be discussed. It is measured in the technology use surveys that have been discussed in the last chapter. It has also been measured in case studies of consumers, an example of which is the kayak community also discussed in the last chapter.

In this section, suggestions are made for how to measure the magnitude of user innovation in a CIS survey. Assuming that the activity is as large as in technology use surveys, there may be a case for doing follow-ups to learn more about user innovation.

There are two ways the user innovator appears in an innovation survey: as a source of an innovation in the form of a prototype or a plan, leading to a product innovation; or as a user of a process which has to be improved or, in the extreme case, developed.

In respect of user innovation of products, there is in CIS surveys a section on sources of information and cooperation for innovation activities. The question, 6.1 in the generic questionnaire, on sources of information includes clients or customers, and it is the high ranking of clients and customers in innovation survey results that establishes them as a key source of information that gives rise to innovation. However, there is no information on whether the information results from service agreements or discussion with sales staff, or whether it consists of a complete set of blueprints or a prototype for a new product.

To probe the role of the user in product innovation, a question could be added, in section 2, after question 2.2 (see Chapter 4) to ask the importance of the user to the product innovation. The following is an example with additions in italics.

> 2.2 Who developed these product innovations?
> Mainly your enterprise or enterprise group
> [*Did the development result from using an earlier version of the product or functionally similar products? (Yes, No)*]
> Your enterprise together with other enterprises or institutions
> [*Were the other enterprises or institutions users of an earlier version of the product or functionally similar products? (Yes, No)*]
> Mainly other enterprises or institutions
> [*Were the other enterprises or institutions users of an earlier version of the product or functionally similar products? (Yes, No)*]

Using the classification established in Table 1.1 and 1.2 in Chapter 1, the first question deals with producer-driven product innovation; the second could be user-driven innovation or user innovation; and the third is user innovation. The 'other enterprise' here could be an individual consumer.

Now consider process innovation. For there to be user innovation in processes, the firm must have to solve a problem related to getting the product it produces to market. The problem could be in production, delivery, market management or in the structure or management practices of the firm. The extreme solution is the development of a new process and the technologies that make it happen; a more moderate case is the purchase of existing technologies, or practices, and then modifying them to do better what is required.

If the firm is able to purchase technologies, or practices, from a supplier and use them without significant change, the firm is not engaged in user innovation as it is not solving a significant problem. It is just updating its capacity to produce and deliver its product by adopting technologies or practices. This act of adoption may result in the firm being classified as innovative if the adoption is within the reference period of the survey and the technology or practices are new to the firm, the lowest level of novelty in the Oslo Manual.

In the technology use surveys discussed in the last chapter, technology was adopted by three means: development; modification; or purchase and use. The first two are user innovation, the third is not.

In the case of process innovations covered in Section 3 of the CIS, question 3.2 could be used as it is to make a first determination of user innovation. The question is repeated here with comments in brackets.

> 3.2   Who developed these process innovations?
>   Mainly your enterprise or enterprise group
> [If the process innovation was developed internally, it would be a case of problem solving needed to move the product to market. This is evidence of user innovation.]
>   Your enterprise together with other enterprises or institutions
> [The work could involve other users or the producer of the process being improved. It could also involve contracting work out to other enterprises or institutions. In both cases, it is the enterprise that is solving its process problem and this is evidence of user innovation.]
>   Mainly other enterprises or institutions
> [This is a case of user adoption and not of user innovation.]

Existing surveys that ask this question show that over half the enterprises developed their own process innovations, and about half that number did so with other enterprises or institutions. The figures from the German survey[1] were (55 per cent, 33 per cent and 12 per cent); Canadian figures for process innovators in manufacturing were (64 per cent, 28 per cent and 7 per cent). This suggests that there is significant user innovation taking place and that there is a case for following this up to learn more about how it is managed and funded and how the intellectual property generated is protected. This has been done for the Canadian survey of use and planned use of advanced technology.

Statistics Canada conducted a survey of the use and planned use of advanced technology in manufacturing for reference year 2007 (AT07) (Statistics Canada 2008b), and then followed-up 1219 responses from plants that adopted technologies by modifying existing ones or by developing them, in the absence of there being suitable technologies on the market (Statistics Canada 2008c). The results appear in Schaan and Uhrbach (2009). Gault and von Hippel (2009) analyse the data with emphasis on the policy implications of the sharing of the intellectual property resulting from adopting by modifying and by developing. The relevant findings are given in Tables 5.1 and 5.2.

The first observation is that developers make more use of conventional intellectual property protection than modifiers, and there will be a size effect here as developers have to be large enough to have the capacity to design, test and develop a process technology. In each of the six categories

Table 5.1   How were user innovations protected?

Responses from innovators that:	Modify existing technologies	Develop new technologies
Does your business unit use any method to protect your process IP?		
Yes	46.4	60.3
No	53.6	39.7
If yes, how do you protect your IP?		
Confidentiality agreements	81.0	85.7
Patents	48.9	64.0
Secrecy	41.5	47.2
Trademarks	29.6	39.9
Copyrights	14.4	22.2
Other	0.7	2.0

Source: Based on Gault and von Hippel (2009), Statistics Canada (2008c).

of intellectual property, developers make more use of them than modifiers, but there are still 40 per cent of developers that do not use existing methods of intellectual property protection and 54 per cent of modifiers.

Table 5.2 addresses sharing of the technologies that have been modified or developed. Between 17 and 19 per cent of firms do share and there are over 200 business units in each category. What is interesting is that 76 per cent of modifiers share their technologies at no charge and an additional 16 per cent share them for some consideration. For developers, there is less free sharing as they already reported making more use of intellectual property (IP) instruments, but 47 per cent of those that share do so for no fee.

The question that arises from Table 5.2 is the presence of economically more efficient sharing of knowledge and how it can be supported by public policy (Gault and von Hippel 2009). Examples are support for 'open licensing' infrastructures such as the Creative Commons license for text and the General Public License for open source software code. Public policy could also support 'defensive publishing' as a mechanism to insure that user innovators, not seeking formal IP protection for themselves, cannot be excluded from using their own inventions and innovations at a later point (Henkel and Pangerl 2008).

Given the evidence from innovation surveys on the undertaking of process innovation in the plant, or with collaborators, there is a case for follow-up surveys similar to that of the Statistics Canada advanced technology survey which would identify the characteristics of the activity and

Table 5.2   How did users share their process innovations? Under what
              terms did sharing take place?

Responses from innovators that:	Modify existing technologies	Develop new technologies
Does your business unit share the technologies that it has modified (developed) with other firms or institutions?		
Yes	17.2	19.0
No	82.8	81.0
How does your business unit share the technologies it has modified (developed)?		
At no charge	75.8	47.3
In exchange for something of value (i.e., free equipment)	16.2	27.7
For a fee	13.1	40.2
Other method	12.1	16.1
Why did your business unit choose to share the technologies that it modified (developed)?		
To allow a supplier to build a more suitable final product	53.9	53.6
Gain feedback and expertise	41.2	48.2
Nothing to lose (no direct competition)	36.3	26.8
Enhance reputation	35.3	46.4
Other	15.7	14.3
Contractual obligation	14.7	28.6

Source: Based on Gault and von Hippel (2009), Statistics Canada (2008c).

provide empirical support for the public policy debate around an intellectual property infrastructure, parallel to the one now in place, that would support the free sharing of knowledge resulting from innovation. The response to the intellectual policy question on the US BRDIS that deals with the free revealing of intellectual property should also contribute to this discussion.

Career paths of doctorate holders
Without people there is no innovation. They make up the markets to which new products are sold and they populate the firms where products

are produced and delivered to the market. The highly qualified are key to the developing, finding and synthesizing of new knowledge, and converting that knowledge to commercial value, and they are in short supply (OECD 2008f). The most highly qualified are doctorate holders, and the understanding of the mobility and the career development of this valuable resource is a challenge in the process of being met.

The case was made in the OECD Blue Sky Forum II (Auriol 2007) and work has been going on since with contributions from the NSF and its work on postdocs and through the use of its survey of earned doctorates. The United Nations Educational, Scientific and Cultural Organization (UNESCO) Institute of Statistics has also been involved. In addition, the subject is being pursued by the EU Commission and by the Organisation for Economic Co-operation and Development (OECD) which has developed a Roadmap for new human resources in science and technology (HRST) indicators to guide the work of member countries.[2]

There is no doubt about the need to know more about the stock and flow of doctorate holders and how their careers develop. Part of the problem lies in the complexity of the data caused by different citizenship practices in different countries, different mobility rules, and different policy agendas that span departments responsible for education, immigration, training, and science, technology and innovation.

The policy question is how to attract and retain the highly qualified researcher in a country that is trying to manage sustainable productivity growth through innovation. This is a subject for Part III. However, it is difficult to develop more effective policies without relevant data.

CARE IN USING INDICATORS

The Single Indicator

. . . does not tell the full story

The use of a single indicator, such as the percentage of GDP devoted to R&D, along with supporting policies, does not provide a full picture of innovation in the country. In the first place, R&D is not innovation and may never result in a new product making it to market, or to the clients of public institutions. Second, in a global economy, it is not clear to which countries the benefits flow that result from the performance of R&D (Freeman and Soete 2007). Finally, as mentioned earlier, not all firms that innovate perform R&D, yet they create value (Rammer et al. 2008). These firms constitute an important, and not well understood, population that contributes to the economy. If the policy objective is to promote the performance of the activity of innovation, more indicators have to be taken into consideration that provide information on firms that innovate, but do not do R&D.

. . . may need another indicator to give it meaning

A single indicator may have to be qualified by the value of another indicator if there are to be meaningful results. The example is the presence of knowledge management practices and the size of the firm (Earl and Gault 2003). As already discussed, the mix of practices engaged in by the firm changes as size changes.

. . . may have to be combined with another indicator

A single indicator may have to be combined with another indicator to provide internationally comparable results. Arundel noted the difficulties in comparing the results of the total sales of products that were new to the market, which was partially rectified by combining that indicator with information on the firm's market, whether it is local, national or international. The combination of an international market and the sales of products provided more credible comparisons (Arundel 2007: 54).

. . . may give different results if it comes from a cross-sectional or panel survey

As mentioned in the last chapter, indicators can be based on information from cross-sectional surveys or panel surveys and both sources can be augmented by administrative data. McDaniel (2006: 162) provides examples of different inferences from repeated cross-sectional and panel surveys. One such example in a social survey is that analysis based on cross-sectional data showed higher incidences of poverty than was found from longitudinal analyses of people over time. The reason was the unevenness of people's lives, with episodes of poverty that came and went, but over time not all were poor. The cross-sectional surveys saw the poverty episodes of people who were not poor and gave rise to inflated population estimates. The same situation could apply to firms that were not innovative in a particular reference period, but were over time. Understanding this behaviour, whether it is poverty or innovation, has significant implications for any policy intervention intended to change behaviour.

Changing Definitions

Statistical measurement evolves and Chapter 3 provided an example of how the definitions of innovation and the coverage of innovation surveys have changed over time. As a result, the propensity to innovate statistic may, in 2009, have a different meaning than it had in 1992. This difference is the four components of the definition of innovation in the 2005 Oslo Manual compared with two in the 1992 first edition.

In some cases, a change in the definition of employment, or of what an institution of higher education might be, could result in statistics that could change the ranking of the country in international comparisons. The UN Fundamental Principles of Official Statistics, adopted by the UN Statistical Commission in 1994, provide guidance on how official statistics, and indicators, should be produced (Box 5.1).

Inappropriate Use

R&D performance is not an indicator of innovation as there is no link in the statistic to the market. Patent counts are not indictors of innovation as there is no information in the counts about the commercialization of the invention protected by the patent. Yet, both have been used as indicators of progress in innovation strategies. They are, of course, correlated with the activity of innovation, but that correlation will be dependent on the size of firm. Small and medium-sized enterprises (SMEs) may innovate to survive, but they will have a much lower propensity to do R&D or to patent.

THE CAPACITY TO USE INDICATORS

There is no point in developing indicators of innovation if they are not used, and used appropriately, as part of the policy process. This includes informing public policy debates, such as whether to conduct research in genetically modified foods, to change the production process as a consequence of this research, and to offer new products to the domestic and the export markets. It also includes informing the discussion within policy departments.

Public policy debate requires participants able to understand the issues and to argue the case for resource allocation in competing domains of innovation. Such debate can take place in a parliament, in universities or research establishments, or in forums run by interest groups. Understanding the issues requires some appreciation of innovation and society, and the capacity to balance the quantitative and qualitative parts of the discussion. This has implications for university curricula, as well as for the institutions that support public policy debate.

Policy departments are there to provide policy advice and to prepare legislation which will guide the country. In dealing with innovation policy,

BOX 5.1   UN FUNDAMENTAL PRINCIPLES OF OFFICIAL STATISTICS

- Official statistics provide an indispensable element in the information system of a democratic society, serving the government, the economy and the public with data about the economic, demographic, social and environmental situation. To this end, official statistics that meet the test of practical utility are to be compiled and made available on an impartial basis by official statistical agencies to honour citizens' entitlement to public information.
- To retain trust in official statistics, the statistical agencies need to decide according to strictly professional considerations, including scientific principles and professional ethics, on the methods and procedures for the collection, processing, storage and presentation of statistical data.
- To facilitate a correct interpretation of the data, the statistical agencies are to present information according to scientific standards on the sources, methods and procedures of the statistics.
- The statistical agencies are entitled to comment on erroneous interpretation and misuse of statistics.
- Data for statistical purposes may be drawn from all types of sources, be they statistical surveys or administrative records. Statistical agencies are to choose the source with regard to quality, timeliness, costs and the burden on respondents.
- Individual data collected by statistical agencies for statistical compilation, whether they refer to natural or legal persons, are to be strictly confidential and used exclusively for statistical purposes.
- The laws, regulations and measures under which the statistical systems operate are to be made public.
- Coordination among statistical agencies within countries is essential to achieve consistency and efficiency in the statistical system.
- The use by statistical agencies in each country of international concepts, classifications and methods promotes the consistency and efficiency of statistical systems at all official levels.
- Bilateral and multilateral cooperation in statistics contributes to the improvement of systems of official statistics in all countries.

Source: UN Statistical Commission (1994).

staff in policy departments must have an understanding of the innovation system, and that rarely does one policy in one department lead to the achievement of the economic and social objectives of the government. Other departments have to be involved, and in most governments this presents a coordination challenge, which is addressed in Chapter 8. Here, the point to be made is that the policy analysts, at the middle and senior levels, should be both literate and numerate, able to put a case using innovation indicators and able to understand the need to fill gaps in existing systems of indicators. Not only should the analysts have such a skill set, but they also require some knowledge of the subject, informed by past experience of initiatives in the department that have succeeded or failed. It is in this environment that monitoring, benchmarking and evaluation lead to policy learning and to more effective policies. How the skill sets and the knowledge are developed is a challenge for government departments, but without them there is no informed demand for the development and use of innovation indicators, and there is no informed policy.

The importance of capacity building in developing countries will appear in Chapter 9, but the last paragraph illustrates a point made earlier in the book that this subject of innovation policy measurement and learning varies by degree from country to country. Capacity building is important and it is a need not just in developing countries.

As a final observation on capacity building, it is not just the policy analyst who has to understand the indicators and how to use them; the official statistician has to understand the problems faced by the policy analyst so that indicators that are timely and useful can be produced and inserted into the policy process. While official statisticians must maintain a distance from government, that does not mean that they should ignore the reasons for which the statistics are being produced.

SUMMARY

This chapter has provided a brief overview of the use of indicators to support policy learning through monitoring, benchmarking and evaluating of innovation policies. Foresight also has a role to play in policy learning.

Indicators also provide information on firms and the institutions to which they are linked, including users of products. This information supports the development of policy.

Indicators can also be used to make a case to develop better indicators to support more effective policy. The two examples in the chapter are user innovation, including the role of the end user as a product innovator; and the career path of doctorate holders, so that there is better information on the mobility and career development of this valuable and expensive resource. Other topics could have been chosen, depending upon the priorities of government. Examples are more statistics on the production and use of new materials, nanotechnologies, biotechnologies, or the user-created content appearing on information and communication technology (ICT) platforms.

Indicators must be used with care and single indicators do not always tell the story that the inexperienced policy analyst, or general user, thinks that they should. Statistics and indicators can also be abused if their creation is not kept distant from the policy process and that is why the UN Fundamental Principles of Official Statistics appear in the chapter, to serve as a reminder.

The importance of the capacity of the policy analysts to use and understand indicators was stressed, as well as the capacity of official statisticians to understand the policy process and where they fit in it. In a global and rapidly changing world innovation activities are interconnected and are influenced by framework conditions in countries. To understand what is going on sufficiently to have influence requires the systems approach presented in Chapter 2. The understanding is also supported by the indicators and their applications developed in this chapter.

NOTES

1. The author is grateful to Christian Rammer from the Centre for European Economic Research (ZEW) for providing this information.
2. The OECD work may be followed on the Career Paths for Doctorate Holders website: www.oecd.org/sti/cdh.

PART III

Innovation strategies

6. Innovation strategies, advice and direction

INTRODUCTION

In Part III, Chapter 6 looks at how innovation is promoted at the international level through the Innovation Strategy of the Organisation for Economic Co-operation and Development (OECD) and then at how it is done through the Innovation Strategy developed by the Commission of the European Communities, the executive arm of the European Union. The approaches of the two organizations are quite different but the objectives are similar: to improve the economic and social well-being of citizens of member countries in a global economy undergoing a financial crisis.

After the international and supranational approaches have been considered, Chapter 7 provides a description of possible components of an innovation strategy. The list of components leads to a discussion of which components should be, or could be, used in an innovation strategy.

What makes a strategy work is the way in which the components are coordinated once the strategy is implemented, and this is a topic for Chapter 8, followed by an analysis of some country experiences in implementing their own innovation strategies. The key observation in Chapter 8 is that there is no single strategy. Not all countries use all possible components, and coordination mechanisms differ for a variety of reasons.

By the end of Part III, the reader should have an appreciation of different approaches to policy, what has been used and what not, and where the work on developing innovation strategies is going in the industrialized economies. The reader should recognize that the overviews of the work of the OECD and of the European Union (EU) in this chapter are just that, and they are written from the perspective of developing innovation strategies, while both organizations do much more. These are complex organizations and there is no claim to completeness. However, the references provide suggestions for further reading.

THE OECD

Mandate and Process

The OECD consists of 30 member countries and its mission is the following:[1]

> OECD brings together the governments of countries committed to democracy- and the market economy from around the world to:
>
> - Support sustainable economic growth
> - Boost employment
> - Raise living standards
> - Maintain financial stability
> - Assist other countries' economic development
> - Contribute to growth in world trade
>
> The Organization provides a setting where governments compare policy experiences, seek answers to common problems, identify good practice and coordinate domestic and international policies.

The original motivation for developing its Innovation Strategy in 2007 was the need to maintain these objectives at a time when world markets were expanding and new players, such as China, India and Brazil, were entering the game. It was an innovation strategy rather than a growth strategy, as innovation was seen as a driver of sustainable economic growth upon which the OECD should be focusing.

The OECD is governed by Council which is normally attended by permanent representatives, but once a year it meets at ministerial level to review activities of the past year and to set its direction for years to come. It was at the 2007 ministerial meeting of Council that the Innovation Strategy was launched.

Once such an initiative begins, position papers are produced by the Secretariat. The Secretariat consists of OECD staff working in support of committees that are attended by senior civil servants from member and observer countries, by the European Commission, and by some international organizations. The OECD is run by the Secretary General, assisted by deputy secretaries general, and the work is done in the directorates which support committees and their working parties. In the case of the Innovation Strategy, it was recognized from the beginning as a cross-cutting activity, but it was overseen by one Deputy Secretary General and coordinated by the Directorate of Science, Technology and Industry (DSTI).

A Learning Organization

The OECD has several advantages as a place to develop an innovation strategy, not the least of which is its reputation for rigorous analysis and the production of guidelines and standards for measuring and interpreting activities which contribute to innovation. In doing both, through its committees, the OECD is a centre of peer learning, peer review and consensus building. The consensus approach means that all member countries are able to accept and support the analysis, the guidelines and the standards. The resulting publications and databases support a culture of lifelong learning. This facilitates the engagement of other players, whether countries or international organizations, and increasingly, private sector stakeholders and civil society. The outcome is an ability to change behaviour, going beyond the presentation and analysis of best practices (Gault and Huttner 2008; Gault 2009).

While the OECD is a platform for peer learning, it is also a learning organization. The measure of this is its ability to change its own behaviour and way of doing business in order to deliver the Innovation Strategy with a minimum of internal transaction costs. In demonstrating that it can do this, the OECD itself provides a case study which illustrates how 'whole of government' policies can be developed across a number of government departments, and implemented. A learning organization that supports peer learning, peer review, consensus building and lifelong learning is able to function effectively because of its accumulation of standards and guidelines that create a common language for the discussion of innovation and its impacts (Chapter 3).

Earlier Projects

Four earlier OECD projects are relevant to the challenge of developing the Innovation Strategy: the Jobs Strategy; the Growth Project; Governing Innovation Systems; and Going for Growth. They have some common characteristics, but stop short of what is needed to deal with a full innovation strategy.

The Jobs Strategy (OECD 1999c) had as a goal the review of policies that gave rise to good jobs at a time when there was high unemployment in many OECD countries. The single objective was complemented by one indicator, unemployment, available in every newspaper in every capital every day. The engaged government departments were few and dealt with labour and unemployment and, perhaps, education policies. The Jobs Strategy of the 1990s was, and still is, a success. However, it required relatively little coordination across the OECD or across departments of governments in member countries.

The Growth Project examined growth in the context of the then 'new economy', dealing with human resource development, technological change and the impact of information and communication technologies on growth. The work was concentrated in one directorate and the objective was not to develop as strategy but to examine determinants of growth. The summary report, *The New Economy: Beyond the Hype: The OECD Growth Project* (OECD 2001d), provided a measured view of the impact of activities in the 'new economy'.

Governing Innovation Systems dealt with governance issues and produced three reports (OECD 2005a, 2005b, 2005c). One of the observations in the study was that countries that performed well in terms of science and technology indicators did not perform well in terms of innovation (OECD 2005a: 29). Lundvall (2007) suggests that this is because of a narrow understanding of innovation. This emphasizes the importance of concepts and definitions (Chapter 3) when conducting a public discourse on the activity of innovation. The work of the project on governing innovation systems, which included work on coordination and policy learning, contributes to Chapter 8. However the reader should keep in mind that the governance project was going on during the three-year revision process of the Oslo Manual, also published in 2005 (OECD/Eurostat 2005). The fact that OECD (2005a) does not cite any edition of the Oslo Manual suggests more a preoccupation with governance mechanisms related to innovation policy than with the activity of innovation.

Going for Growth (OECD 2008g), reviews policies related to economic growth. In Chapter 3 of the 2006 report (OECD 2006a) there is an overview of policies for the encouragement of innovation, and there are suggested indicators related to research and development (R&D) and to patents. It is acknowledged that 'resources devoted to R&D are not sufficient to assess a country's innovation outcome' (OECD 2006a: 59), and additional indicators are proposed with a discussion of their strengths and weaknesses. These are the propensity to innovate by firms, broken down by sector; the share of population aged 25–34 with at least tertiary education; mathematical and scientific literacy of 15-year-olds; venture capital investment; and the share of R&D performed by foreign affiliates.

The annual *Going for Growth* reports, starting from 2005 (OECD 2005d, 2006a, 2007d, 2008g, 2009f), provide analysis and advice on aspects of economic growth, not all of which are related to innovation. The approach fits well with that of Departments of Finance or Economics in OECD member countries, but does not necessarily require the involvement of other government departments.

These four examples show that the OECD can take on major issues – jobs, governance of innovation and growth – and provide relevant and timely analysis and advice. What is missing when it comes to innovation is the need to coordinate across a wide range of OECD directorates, and other groups, in order to deal with all aspects of the innovation system and its behaviour.

There is a fifth example that could be regarded as a precursor of the current Innovation Strategy. It is the Technology Economy Programme (OECD 1992b) which was initiated by Council in 1988, just after the financial crises of October 1987, and which reported in 1991. It addressed the concerns of OECD member countries in a period of major change, for a better understanding of the interactions between technological development, the economy and the society (OECD 1992b: 3). It was a cross-cutting initiative at the OECD involving a number of directorates and was coordinated by the Director of Science, Technology and Industry, Robert Chabbal, with the research managed by François Chesnais. There were contributions from the then Economics and Statistics Department and the Directorates for Social Affairs, Manpower and Education, Environment and Development Co-operation, as well as the Development Centre.

While the world has changed since the period of 1988–91, the 1992 OECD report is still worth reading, and it, and the work done for the conferences that formed part of the project, and the findings of the High Level Group of Experts, continue to influence thinking about innovation research.[2]

THE OECD INNOVATION STRATEGY

In 2007, the OECD Council, meeting at ministerial level, initiated the Innovation Strategy. The following summary presents the rationale and the objectives of the Strategy:

> Ministers concluded that in order to strengthen innovation performance and its contribution to growth, a strategic and comprehensive cross-government policy approach is required. They recognised the OECD's high-quality contributions in the area of innovation and requested that the OECD deepen its work in this domain. They welcomed plans for an OECD Innovation Strategy, along the lines of the OECD Jobs Strategy, which could make an important contribution to policymaking in OECD and non-OECD economies. The Strategy would be formulated around:
>
> - evidence-based analysis and benchmarking;
> - a framework for dialogue and review;
> - new indicators on the innovation–economic performance link;
> - initiatives for innovation-friendly business environments; and
> - the development of best practices and policy recommendations.

The strategy could draw on relevant OECD work on innovation, entrepreneurship and the broader business environment. Ministers particularly welcomed the incorporation of cross-cutting work on innovation to address global challenges, notably in the environmental and health domains, globalisation of innovation, evaluation of innovation policies and country-specific analysis. They asked the OECD to study the impact of innovation on the services sector. The OECD could also examine the functioning of the current IPR [intellectual property rights] system in the context of the new, more open, business environment for innovation and propose ways to ensure an adequate balance between stimulating innovation and providing access to knowledge. The proposal to undertake a project on innovation in the software sector was welcomed as a useful contribution to this effort. (www.oecd.org/mcm2007) (The bullets have been added by the author.)

A progress report was made in 2008 (www.oecd.org/mcm2008) and in 2009 the summary interim report (OECD 2009g) was made to Council which responded:

We also look forward to the results of the OECD Innovation Strategy, as an important source of policy guidance for boosting productivity, competitiveness and growth, and for harnessing innovation to address global challenges. (www.oecd.org/mcm2009)

The focus on growth and global challenges remains, but the context has shifted to innovating out of the financial crisis that altered world economies in 2008–09.

The interim report lays out the changes in innovation in recent years and the policy challenges of increased complexity, shorter timescales, unpredictable dynamics, global reach and the non-linearity of response to policy interventions which were raised in Chapter 1. It then goes on to present the areas in which the policy advice will be forthcoming in the final report to Council in 2010, with emphasis on whole-of-government policies for innovation.

People, Coordination and Soft Skills

In emphasizing the importance of coordination of policy across government departments the interim report reflects the original expectation of ministers in 2007. The report also notes the need to examine the role of coordination and cooperation in innovation activities in the private sector and to see where policy could facilitate such activities. With globalization, innovation has to be coordinated along value chains and within networks. There are knowledge flows to be coordinated between producers and users of the products produced and with suppliers. Knowledge markets also raise coordination issues.

These are people issues, with implications for their education, training and lifelong learning activities. They go beyond coordination to the soft skills required to interact effectively in networks, and to capture knowledge from networks or knowledge markets and to convert that knowledge into value as part of the innovation process. Other people issues are preparation for entrepreneurship and for risk taking. A consequence of the financial crisis is greater risk aversion which has to be countered if there is to be the innovation to support the economic growth needed to recover from the crisis.

People influence markets and that has implications for communication and coordination activities. An example is the aversion of some people towards genetically modified foods, with implications for trade in food produce and for agricultural practices.

Firms and their Characteristics

The interim report raises a number of issues about firms in the innovation system. The importance of intangibles, other than R&D, to the innovation capacity of the firm is one. Entrepreneurial activity is another, and the question of whether policy should focus on the age of the firm rather than, or in addition to, the size of the firm. If new entrepreneurial firms are to grow, they need support in the form of various stages of capital inputs, but also management and technical mentoring. If they are to participate in global markets, they need not just access to information and communication technology (ICT) platforms, but broadband access and mobile telecommunications, and advice on what to do once they have the access.

Non-technological innovation is identified as an issue, including organizational innovation and the use of management practices. If the majority of innovative firms, especially smaller ones, perform little or no R&D, there is a need to understand better how these firms function when they create value for the market. There are also differences between firms in the service sector and those in the manufacturing sector, although neither sector is homogeneous.

Framework Conditions, Learning and Governments

As discussed in Chapter 1, governments provide the framework conditions that support, or inhibit, innovation. The interim report deals with these as key issues and makes reference to the demand for innovation. This suggests that procurement policy could be further developed as part of the innovation strategy, subject of course to competition and trade policy.

The fact that innovation activities are spread over space and over time and across subject matter calls for an integrated approach to policy intervention, although the reality differs from country to country (Chapter 8). There is also a need for policy learning to take place and to be an objective of an innovation strategy and part of its implementation, monitoring and evaluation. Policy learning here is institutional learning, which is different from the learning of individuals but it is at least as important. It brings together the human capital, the network capital to which the people are connected, and the institutional capital and corporate memory of the government department or departments that are managing innovation policy or policies.

Development and Global Challenges

A significant observation of the interim report is the need to bring innovation policy into development policy in a coherent way and to enable value creation through entrepreneurship in developing countries. Agriculture is singled out as a sector where innovation can be a key driver of poverty reduction (OECD 2009h). The promotion of access to mobile communications is seen as a trigger for local innovation to advance rural development beyond agriculture, while recognizing that the improvement of rural productivity requires investment in infrastructure.

Measurement

The interim report takes a broad view of innovation and recognizes that this requires additional collection and use of internationally comparable data at the firm level, as well as a better understanding of currently unmeasured factors in the innovation process needed to understand the complexity of innovation. It also makes the point that evaluation is essential to better policy making, and this also requires better indicators.

Next Steps

As stated in the interim report, 2009 and 2010 are to be spent providing policy guidance, based on the broad principles described in the report, to support the development and implementation of effective whole-of-government policy recommendations for innovation.

THE EUROPEAN UNION

While the OECD is producing an innovation strategy, so also is the European Union. As the structure and objectives of the EU influence the development of its strategy, and its coordination, the section begins with a short description of how the EU functions. Nauwelaers and Wintjes (2008) provides an overview of innovation policy in Europe.

Organization

The European Union is a supranational organization which, according to the EU website (http://europa.eu/abc/12lessons/index_en.htm), is more than a federation of the 27 member countries, but not a federal state. It is a new structure. The website provides a description of how the EU works:

- The Council of the European Union, which represents the member states, is the EU's main decision-taking body. When it meets at heads of state or government level, it becomes the European Council, whose role is to provide the EU with political impetus on key issues.
- The European Parliament, which represents the people, shares legislative and budgetary power with the Council of the European Union.
- The European Commission, which represents the common interest of the EU, is the main executive body. It has the right to propose legislation and ensures that EU policies are properly implemented.

It also makes clear that the priority of the EU is on growth and jobs:

- The Union intends to respond to globalization by making the European economy more competitive (liberalization of telecommunications, services and energy).
- The Union is supporting the reform programmes of member countries by making it easier to exchange 'best practice'.
- It seeks to match the need for growth and competitiveness with the goals of social cohesion and sustainable development which are at the heart of the European model.
- The EU Structural Funds will spend more on training, innovation and research in the 2007–13 budget period.

It is within this decision-making and operating structure and with the priorities just presented that the EU Innovation Strategy emerged from the European Commission, also known as 'the Commission' or the 'Commission of the European Communities' (CEC).

The EU develops and implements policy. In doing this, Expert Groups are called into being when needed, to complement the knowledge of Commission staff. Policy learning results from the process of creation, approval and implementation of policy. Once the policy is implemented it is monitored and evaluated and its impacts are observed. At the EU level, the policies are expected to demonstrate 'additionality' when compared with the policies of individual European countries, such as those discussed in Chapter 8. Examples of EU-level policies are the Seventh Framework Programme for Research and Development (Muldur et al. 2006), the Competitiveness and Innovation Programme, and the Cohesion Policy. As with the OECD, the EU also has a separate Growth and Jobs Strategy (CEC 2007a) and it established a Competitiveness and Innovation Framework Programme for 2007–13 in 2005 (CEC 2005).

EU INNOVATION STRATEGY

The EU Innovation Strategy is set out in the Commission document CEC (2006b), and it contains the components of an innovation strategy and the need to coordinate through the engagement of all parties – business, public sector and consumers. The document also makes the strong statement that: 'Europe does not need new commitments; it needs *political leadership and decisive action*'. It ends with ten actions of high political priority as part of the Lisbon Strategy for growth and jobs. As an EU document it addresses the benefits of removing barriers to the working of the internal market, with special attention to the service sector and the improvement of the institutional framework for European standardization to ensure global success for European companies. The ten action items are summarized in MEMO/06/325 (www.europa.eu/rapid):

1. Establish innovation-friendly education systems
2. Establish a European Institute of Technology
3. Work towards a single and attractive labour market for researchers
4. Strengthen research–industry links
5. Foster regional innovation through the new cohesion policy programmes
6. Reform R&D and innovation state aid rules and provide better guidance for R&D tax incentives
7. Enhance intellectual property rights protection (IPR)
8. Digital products and services – initiative on copyright levies
9. Develop a strategy for innovation friendly 'lead markets'
10. Stimulate innovation through procurement.

The document recognizes four directions for future work. They are the need to:

- understand the specificities of innovation in services, a measurement issue;
- support all forms of innovation, not only technological innovation;
- develop specific support mechanisms for innovative services with high growth potential; and
- foster transnational cooperation on better policies in support of innovation in services in Europe.

Nine Priorities from the Competitiveness Council

After the release of CEC (2006b) the Competitiveness Council identified nine priorities for innovation action at the EU level. They were the following:

1. An intellectual property rights framework.
2. Standardization in support of innovation.
3. Public procurement in support of innovation.
4. Joint Technology Initiatives (JTIs).
5. Boosting innovation and growth in lead markets.
6. Enhancing closer co-operation between education, research and business by establishing the European Institute of Innovation and Technology (EIT).
7. Regional innovation through cluster promotion.
8. Innovation in services.
9. Facilitating risk-capital markets.

Not surprisingly, the Competitiveness Council has focused on market-related activities. The IP, procurement, lead markets, the EIT, regional innovation and risk capital are priorities in common with those of the broad-based innovation strategy (CEC 2006b), while standards, JTIs and services are found in the body of the Innovation Strategy document. Those present in the broad-based Innovation Strategy list, but not present in that of the Competitiveness Council, are public sector priorities such as education reform in the context of the Bologna Process, supporting mobility for researchers, and the reform of rules for state support of R&D and innovation.

Intellectual property rights

The Commission has published *An Industrial Property Rights Strategy for Europe* (CEC 2008c) which recognizes the need for a clear regime for intellectual property rights as a condition for the single market and support for the 'fifth freedom', the free movement of knowledge (the other four are the free movement of people, goods, services and capital). The strategy document goes beyond patents to include copyrights and trademarks and other means of IP registration, and it includes a section on development issues and the EU (see Chapter 9). Prior to the strategy was a Pro Inno Europe Memorandum (Pro Inno Europe 2007a) on removing barriers for a better use of IPR by small and medium-sized enterprises (SMEs).

Standardization in support of innovation
Standardization is seen as having the power to accelerate the access of innovation to both domestic and global markets (CEC 2008d). It also has the power to inhibit the uptake of innovation if standards are not available, are out of date or are contradictory. The objective of improving standardization in the EU is to complement market-based competition and to make it easier for innovation to happen and for the results to be diffused. It is seen as providing a way to gain first-mover advantage in global markets and, therefore, is a competitiveness tool as well as an innovation tool. For this reason, standards appear in the EU's Growth and Jobs Strategy. They also appear in the lead market initiative and in public procurement policy.

Public procurement in support of innovation
Public procurement is being recognized as a way of supporting innovation which also links to the lead market and the standardization initiatives. Government can be a critical user in its approach to procurement and it can also influence innovation through its use of standards as part of the procurement process. The Commission has released a guide on dealing with innovative solutions in public procurement (Pro Inno Europe 2007b).

Edler and Georghiou (2007) provide some background to the development of procurement policy in Europe, making the point that procurement policy, as a means of stimulating innovation, is more efficient than a wide range of R&D subsidies. They situate public procurement in the context of systemic public policy, regulation and the support of private demand, and argue that for best effect all aspects of public policy should be used to advance innovation. This emphasizes the coherent approach to innovation policy.

Joint Technology Initiatives (JTIs)
Joint Technology Initiatives (JTIs) address a number of objectives as part of the broad-based Innovation Strategy. They address a market failure and promote the development of technologies that are consistent with European priorities. Current examples are: embedded computer systems; innovative medicines; aerospace (Clean Skies); nanoelectronics; hydrogen and fuel cells; and the Global Monitoring for Environment and Security (GMES) initiative with the European Space Agency (ESA). The means of promoting the development of technologies is through public–private

partnerships, including the Commission, a not-for-profit industry-led association and, possibly, member states. The support for the JTIs comes from the Seventh Framework Programme with possible additional funding from the European Investment Bank. The public–private partnership ensures a critical mass of researchers and a connection to the market, and the technologies ideally are chosen to support competitiveness.

Lead Market Initiative (LMI)
The Lead Market Initiative (LMI), according to the EU website (ec. europa.eu), is one of the most important innovation policies in the EU, involving member states, industry, non-governmental organizations (NGOs) and the European Commission. It involves a set of core policy instruments, including legislation giving rise to policy instruments, public procurement policy, standardization, labelling and certification policy, and complementary instruments such as financial support and incentives to facilitate the interaction of customers and innovating firms. As a consequence, the LMI is cross-cutting and requires the involvement of various parts of the Commission, and member states.

The Commission definition (CEC 2006b) of a lead market is: 'where an innovation is first widely used that later becomes successful internationally regardless of where that innovation was invented'. Blind et al. (2009) in their paper on the monitoring and evaluation methodology for the LMI review other definitions and means of evaluating a lead market initiative.

Six lead market areas have been selected. They are eHealth; sustainable construction; technical textiles for intelligent personal protective clothing and equipment; bio-based products; recycling; and renewable energy (CEC 2007b). A mid-term progress report was planned for 2009 and a final report in 2011, including evaluation and an assessment of the impact of the policy actions and, where possible, the impact of the lead markets.

European Institute of Innovation and Technology (EIT)
The EIT has been established and has issued its first call for proposals from experts to assist the EIT in connection with the evaluation and implementation of Knowledge and Innovation Communities (KICs). This is a step towards achieving its goals of promoting interactions between research institutions and industry, supporting knowledge transfer and bringing about added value to existing EU initiatives (eit.europa.eu).

Clusters
In 2008 the Commission adopted a Communication, *Towards World-Class Clusters in the European Union: Implementing the Broad-Based Innovation Strategy* (CEC 2008a) and its annex with concepts, definitions and

statistics (CEC 2008b). A Commission Decision (OJ L 288/7 22.10.2008), established the European Cluster Policy Group. Prior to that, a European Cluster Observatory was established in 2007 to provide information about cluster policies in 32 countries. The objective of all of this was to support the emergence of world-class clusters in the EU.

Innovation in services
Innovation in services gave rise to a Commission Staff Working Document in 2007 (CEC 2007c) and a call for proposals in May 2007 to establish a European Knowledge-Intensive Services Innovation Platform (KIS-IP).

Risk capital markets
In December 2007, the Commission adopted a Communication on *Removing Obstacles to cross-Border Investments by Venture Capital Funds* (CEC 2007d). The Communication addresses the free movement of capital, the improvement of conditions for institutional investors such as pension funds to provide venture capital in member states, the improvement of regulatory frameworks, the reduction of tax obstacles and progress towards mutual recognition of existing national frameworks. The last point would permit venture capital funds in one state to be recognized by other states, avoiding a requirement to be registered in all states where the fund was active.

Six Other Topics

In addition to the nine priorities of the Competitiveness Council, there are six topics on which work is being done:

- Regional innovation.
- Design and creativity.
- State aid.
- Knowledge and technology transfer.
- Eco-innovation.
- Skills.

Regional innovation
Regional innovation is linked to the clusters priority as well as to cohesion policy. Efforts to strengthen the research potential of the regions are supported by the Seventh Framework Programme.

Design and creativity
The Commission links user-driven and user-centred innovation with design and creativity and has produced a working document on Design as

a Driver of User Centered Innovation[3] (CEC 2009a). This has given rise to a public consultation on what should be done by the EU in the area of design.

The EU also designated 2009 as the European Year of Creativity and Innovation, based on a premise that creativity is the prime source of innovation. Creative skills are seen as necessary to tackle the global challenges of climate change, poverty and the consequences of globalization.

State aid
State aid deals with those measures that are compatible with the European common market and it is overseen by the Directorate General for Competition. One of the instruments for promoting innovation through state aid is described in the Community Framework for State Aid for Research and Development and Innovation (OJ C 2006: 26).

Knowledge and technology transfer
This initiative recognizes that competitiveness requires not just the creation of new knowledge but also its conversion into new goods and services. To this end the Commission has recognized the importance of transferring knowledge between public research institutions and third parties, including industry and civil society organizations.

Eco-innovation
Eco-innovation is linked to the Lead Market Initiative, to sustainable industrial policies and to the work of the Directorate General for Energy and Transport. It involves process management (environmental R&D and waste management and recycling) and resource management (water supply, recycled materials, renewable energy production, nature protection and eco-construction). The work of the Commission is described in the *Action Plan on the Sustainable Production and Consumption and Sustainable Industrial Policy* (CEC 2008e, 2009b).

Skills
The Commission recognizes that the skills required for innovation go beyond those needed for science, technology and engineering, and includes those needed for commercialization, management, design, organization, marketing and finance activities. As education, training and lifelong learning are key components of skill development and maintenance, the Directorate-General for Education and Culture is involved as are programmes for e-skills and entrepreneurship.

European Innovation Plan

The European Council, at its meeting in December 2008, called for a European Plan for Innovation. The Commission will present short-term actions in response to the economic crisis, an assessment of the 2006 Broad Based Innovation Strategy and reviews of the Lead Market Initiative, Innovation in Services, financing innovation in SMEs and the effectiveness of innovation support measures.

These activities were expected to provide input to a European plan for innovation to be presented to Council by the end of 2009. The work on the innovation plan will be part of reflections on the Lisbon Strategy, post-2010. In July 2009, the CEC produced a Communication (CEC 2009c), on *Reviewing Community Innovation Policy in a Changing World*. This document reported on the work and the progress of the Commission in the area of innovation policies and noted that innovation and entrepreneurship are not yet sufficiently recognized as values everywhere in Europe and failure was stigmatized, rather than being seen as part of learning. Public sector procurement was seen as a potential driver of innovation, and there was a need for reformed framework conditions that reward innovation in the single European market. The slow start of the Lead Market Initiative was attributed to a lack of synergy between policies and instruments at different levels across the EU. This led to consideration of coordination of innovation policies.

The report states that: 'innovation support involves seven different Commission services, various agencies and 20 committees with representatives from Member States', and calls for 'clear structures and substantial simplification of participation rules for all innovation funding, regardless of its origin'. The coordination and simplification challenges are not peculiar to the Commission, or to the OECD, but they are of fundamental importance. They will be discussed further in Chapter 8.

The report ends with a proposal to explore with member states, before spring 2010, the feasibility of a 'European Innovation Act encompassing all the conditions for sustainable development and which would form an integral and crucial part of the future European reform agenda'.

SUMMARY

The OECD and the EU are different in that the OECD provides policy advice to countries, which may or may not be taken, while the Commission is able to propose policy and to implement policy approved by Council and the Parliament. That means that the EU innovation policy builds on past policies and has to work across the directorates general of the Commission. The OECD is less constrained, but also has to work across its own directorates.

Both organizations, in 2009, were preoccupied with the financial crisis and ways in which innovation policy can help restore the global economy. In both organizations, 2010 is a critical year. It is the end point of the Lisbon Strategy, the start of the EU innovation plan, and the delivery year for the OECD Innovation Strategy and, possibly, the European Innovation Act. It is also a time when a US Innovation Strategy has been announced (Executive Office of the President 2009) and this is likely to have repercussions for both the OECD and the EU strategies.

Neither the EU Innovation Strategy activities reviewed in this chapter, nor the OECD Interim report of the Innovation Strategy, address in any detail a number of questions related to developing countries, such as genetically modified (GM) foods and the impact of banning imports of GM foods. The ban is not just a European issue as it has resulted in similar bans in African countries which need the European market to survive (Collier 2008). However, in the OECD interim report, there is material on development and the importance of getting innovation onto the development agenda. There is also discussion of global challenges and the need to address them through innovation, including green innovation (OECD 2009i).

The role of the user in innovation is recognized in both the EU and the OECD strategies, but more from the perspective of the user-driven[4] innovation defined in Chapter 1. The importance of user innovation, discussed in Chapter 5, is not well developed by either the EU or the OECD at the time of writing in 2009. From a policy perspective, as discussed earlier, promoting user innovation, and innovation not driven by R&D, is a path to firm growth. As larger firms have a higher propensity to do R&D such promotion policies could lead to the performance of more R&D by firms.

NOTES

1. The mission of the OECD can be found at www.oecd.org.
2. Consider substituting the word 'innovation' for the word 'technology' if OECD (1992b) is reviewed.
3. 'User-centered' is one of the terms describing user involvement in the innovation process. The term used in other sources, and the preferred term in this text is 'user-driven' innovation. See Tables 1.1 and 1.2 in Chapter 1.
4. The Commission is actively promoting user-driven innovation and has released a paper (Pro Inno Europe 2009b) on fostering user-driven innovation through clusters as an input to meetings and discussions leading to the European action plan for innovation / European Innovation Act (CEC 2009b).

7. Innovation strategy components

INTRODUCTION

Innovation is constrained or advanced by the cultural, geographical, and legislative and regulatory environment in which it happens. An innovation strategy, if it is to be effective, has to take account of these conditions to ensure that any interventions combine to contribute to the policy goals and do not weaken one another. This raises questions about which policy goals are to be addressed, and then how the activities in their support are coordinated. The focus of this chapter is on the activities that could contribute to an innovation strategy.

KEY COMPONENTS

Potential key components of innovation strategies are grouped under six headings: markets, people, innovation activities, public institutions, international engagement and global challenges. Not all the components turn up in all strategies but the objective is to present them and raise questions about how policies can advance the objectives served by the components.

The topics discussed in this section have a wide range of applicability. Some are appropriate to the developed countries and others can be used in developing countries. They could be classified differently, as healthcare or education could be provided by the private sector as well as the public sector or through a public–private partnership. This is a question for the practitioners of the science of innovation policy. In Chapter 8, the issue is how to bring all or some of these topics together under an innovation policy umbrella and to ask if this is possible or even desirable.

Markets

Brand recognition
Presenting the country as the best place on earth to live and work, to do R&D, to innovate, to manage trade, and to enjoy a high quality of life, supported by a first-class infrastructure, in a safe and attractive environment,

is a goal. A country that can establish and maintain such a brand can attract highly qualified people, foreign direct and portfolio investment, and retain the inward flow during times of economic and social turmoil. Being the best place to be also has implications for the education, training and development of the people who support the infrastructure and provide the non-tradable services (LO 2008).

Some countries would argue that their 'brand' is sufficiently well established and that this component is not necessary. The objective of providing a list of components is not to promote them, but to provide the opportunity to reject, accept or add to them based on rational choice.

Lead market
A highly educated population with intellectual curiosity could be a lead market for technologies and for applications that use the technologies. Lead markets are attractive to leading-edge producers of goods and services, but there is a danger, pointed out by Christensen, of listening only to the most advanced customers (Christensen 1997, 2008). Governments, through procurement and support for trade, can contribute to the lead market.

The European Union (EU) has launched a Lead Market Initiative (LMI) which is an important innovation policy, discussed in Chapter 6. It focuses on six markets: eHealth; sustainable construction; technical textiles for intelligent personal protective clothing and equipment; bio-based products; recycling; and renewable energy.

Competitive engagement
One of the reasons that the country is the best place to do business, to create knowledge and to live is that it supports an outward-looking approach to business. This includes the capacity to participate in and manage global value chains and a culture that supports the learning of languages and international involvement. The goal is to be an effective player on the international stage. This requires outward-looking people with the skills needed to play in the international arena and it has implications for education and training and for cultural institutions.

Financial services
Firms require finance to start up, to do research to produce new products, and to bring the new products to market. While there are banks and other financial institutions that can support established firms, there is a need for intelligent and patient angel investors and venture capital firms that understand the sector, the market and the risk of trying to bring new products to the market.

People

Labour force

People are part of the means of production, and in a global economy trading in knowledge products the workers have to be well educated, self-directed and able to engage in lifelong learning. As part of global engagement, some experience in their career gained outside of the country can be considered an asset. These requirements have implications for education policy and reform, training and development policy in public and private institutions, and migration policies that encourage the mobility of the highly qualified (OECD 2008f).

However, the labour force is not just made up of the highly qualified. There are many more people that are part of the economy and society who produce goods and services, tradable and non-tradable. Their approach to these tasks is part of making the country the best place to be in. Education and training policies must take account of the needs of the entire workforce, as well as the priorities of government.

Demographics and demand for innovation

People are a source of opportunities for an innovation strategy. In most of the industrialized countries, the population is ageing. This is a technical and organizational opportunity to care for an ageing population and to gain new and marketable knowledge from this activity. Also, people embody knowledge, and as their departure from the workplace accelerates, there is a need to capture and retain the knowledge that is being lost. This is an opportunity for non-technological innovation using the techniques of, for example, knowledge management.

Migration

With globalization, the highly qualified are becoming more mobile and Organisation for Economic Co-operation and Development (OECD) countries are net beneficiaries of this (OECD 2008f). This has implications for innovation policies as the highly qualified contribute to the creation and diffusion of knowledge (Auriol 2007). From the perspective of the sending countries, there are issues about using the diasporas (Kuznetsov 2006) as a source of knowledge and of remittances. When it comes to the impact of mobility on innovation there is little or no evidence (OECD 2008f), and this is one of the statistical challenges for innovation policy. There is also a question of how mobility policies fit into innovation policies and the extent to which intervention in these two areas are coordinated.

Innovation Activities

Technologies and practices
Firms can innovate by adopting technologies new to the firm. Governments can provide incentives to do this, especially if there is a national view of which technologies to support. In the German High-Tech Strategy (BMBF 2006), 17 technologies and practices are advanced.

User innovation
Information and communication technologies (ICTs), and to a growing extent biotechnologies, are modular platform technologies that provide a basis for innovation on the platform. The platforms also make it easier to modify the technologies to suit user needs and to create knowledge in the process. This user innovation has always been present in the economy (von Hippel 1988), but now it is easier (Dyson 2007) and it raises questions about how the intellectual property created by the activity is managed (Gault and von Hippel 2009; von Hippel 2005). Consumers can also engage in user innovation by modifying a product to serve their needs and then presenting a firm with the prototype or blueprints to produce the product commercially.

User-driven innovation
User-driven innovation describes the exchange of information between a user of a product and the producer, with a view to improving the product. It does not involve the transfer of prototypes or blueprints. This flow of information is a well-established input to the innovation process.

Open innovation
While ICT platforms encourage user-initiated innovation, they also enable the flow of knowledge across the boundaries of countries and institutions resulting in more 'open' innovation (Chesbrough 2003; OECD 2008h). This takes various forms, of which the open source approach in software development is one, but there is also the drawing of new ideas and technologies into the firm and the outsourcing of activities. The walls of the firm are porous, but a consequence of this is the requirement for people to be able to work with the international networks that are readily available, in addition to more local networks where participants can meet face to face. The expanded use of networks means that knowledge is not just stored in people or embodied in machines and practices, but it is also stored in the network. People can do more things, and more things better, because of the network capital that they can draw upon and contribute to. Enhancing network capital is a goal for an innovation strategy. Measuring it is another matter.

Demand-driven innovation

From the perspective of the firm, demand-driven innovation is a response to the procurement requirements of other firms and governments. This is considered again under the 'Procurement' component in the next subsection.

Public Institutions

Public institutions set priorities, educate and develop the workforce, do research, manage the public healthcare system, and set policies that govern mobility. These activities can influence innovation and, in doing all of these things, public sector institutions can engage in the same innovation activities as go on in the private sector. Here, the activities of public institutions that could form part of an innovation policy are presented.

Infrastructure

Technology and practices provide the infrastructure that supports the economy and the society. The infrastructure includes the information and communications technology (ICT) networks, and well-managed roads, ports and logistic services. Technologies and practices are also integral parts of the education, research, health and financial services infrastructure. A first-class infrastructure is an important element of an innovation strategy. While the components of the infrastructure may be in place in OECD countries, no policy-maker would argue that the infrastructure works as well as it should, or does not need reform.

Procurement

Governments at all levels, education and research institutions, and health institutions have enormous purchasing power which can be used to influence the development path of technologies and practices. However, it is not just the purchasing power that matters but the leadership in the procurement. A well-established example is the case of numerically controlled machine tools (Mansfield 1968: 111). These tools were developed by the Massachusetts Institute of Technology on a contract from the US Air Force, and appeared in 1951. They were then commercialized by the industry and introduced in 1955. The use of these machine tools by industry then allowed the US Air Force to use numeric specifications in its procurement, with impact on all procurement of parts made by machine tools.

The procurement process allows public institutions to be lead users and to provide critical feedback to the suppliers. Extreme examples of this arise in the case of large scientific establishments which are pushing the frontiers of the possible and need computing and measurement speeds, analytical capacity and materials that do not exist in the commercial world. Solving the scientific and engineering problems produces knowledge that can be commercialized.

In a global world, with freer trade, it is more difficult to use procurement to develop the domestic market, but the role of procurement in innovation policy is a key issue which has been neglected (Edler and Georghiou 2007). However, it is part of the EU Lead Market Initiative. It is also an objective of the UK Small Business Research Initiative (SBRI)[1] which aims to use government procurement to drive innovation. It provides business opportunities for innovative companies while solving the needs of government departments.

Priority setting

Expertise in ICTs, biotechnology, nanotechnologies, new materials, energy sources, and other technologies and applications require highly qualified but scarce human resources, and this raises a question as to whether an innovation strategy should involve priority-setting in order to make the most of the knowledge available in the country. As sectors differ significantly in their requirements, this has to be taken into account.

As examples, reference has been made to the German list of 17 practices and technologies (BMBF 2006). The US has identified 'four practical challenges':

- applying science and technology strategies to drive economic recovery, job creation, and economic growth;
- promoting innovative energy technologies to reduce dependence on energy imports and mitigate the impact of climate change while creating green jobs and new businesses;
- applying biomedical science and information technology to help Americans live longer, healthier lives while reducing health care costs; and,
- assuring we have the technologies needed to protect our troops, citizens, and national interests, including those needed to verify arms control and non-proliferation agreements essential to our security. (OMB/OSTP 2009)

Canada has four strategic research areas:

- environmental science and technologies;
- natural resources and energy;
- health and related life sciences and technologies; and
- information and communication technologies (Government of Canada 2007).

Standard-setting

Standards are integral to innovation policy and to trade policy. They can be set by international organizations, such as the International Organization for Standardization (ISO), a specific example being the ISO/TC 229 work on nanotechnology standards, or they can evolve from use and become de facto industry standards. They can deal with technologies and also with how research is done, an example being bioethics and related standards of practice. The European Commission summarizes its position as follows: 'the global promotion of EU norms and standards and innovative initiatives can give a decisive first mover advantage to European companies in the spirit of the lead market initiative' (CEC 2006b: 6). Standard-setting is part of the EU Lead Market Initiative.

Public finance

If new firms are to be created, and to survive and grow, they need finance at various stages of their development and that includes angel investors, venture capital, and support from development banks and from the established banking system. The public sector provides a regulatory environment that maintains confidence in the system while allowing it to provide the services needed, both national and international. It may also provide development banks to fill gaps not covered by the private sector, and export development banks to support firms in the export of their goods and services.

In addition, departments of finance can stimulate innovation, and its components, through tax policy, such as research and development (R&D) tax credits and capital consumption allowances adjusted to encourage capital investment in particular technologies. There are those who would argue that innovation policy *is* tax policy (Licht 2008).

Government direct support

Departments of government spend significant amounts on targeted support programmes, such as the US Small Business Innovation Research (SBIR) programme or the Canadian National Research Council Industrial Research Assistance Program (NRC-IRAP).[2] They also provide direct support for research and development through grants, contracts and contributions, and through mission-related research and collaboration with researchers from business and higher education.

Government departments can also promote dialogue with society on issues that can affect markets for new products, such as genetically modified foods, working conditions in countries from which products are imported, or regulation of financial service industry products. Such dialogues can also encourage the culture of innovation.

Education, training and research

The institutions, public and private, are challenged to produce numerate and literate people capable of assessing the risks resulting from innovative activity, and the rewards. How this is done raises issues of reform of the institutions providing the education and training. However, monitoring the place of education in the innovation system has raised questions about what is being measured and what the consequences might be of producing a misleading set of indicators (Hawkins et al. 2007).

Knowledge is another product of institutions of education and research, and the issue is how this knowledge is protected through the use of intellectual property instruments and then how it is commercialized. The same question arises with government laboratories.

Health

Health institutions have opportunities to be innovative in providing health services, and some of this responds to the impact of private sector innovation. An example of private sector innovation is the provision of standardized foods that contained transfats and sugar leading to obesity and Type II diabetes, putting pressure on the scarce resources of the healthcare system. This is similar to the financial service example used earlier where private sector innovation resulted in unexpected outcomes which placed a demand on public resources. The difference is the time scale; the first was measured in months, the second in years.

Health institutions do research as well as provide services, and there is the broader issue of justifying the expenditure on research in public institutions in the health sector (Bernstein et al. 2007).

Monitoring and evaluation

Innovation strategies are conceived by governments, ideally in consultation with stakeholders, as a means of achieving goals of importance to the country, such as sustainable growth or mitigation of climate change. Then they are implemented, using a selection of the components given in this chapter, or others. Once the strategies are implemented their activities must be monitored and progress towards the goals evaluated. Without monitoring and evaluation, there is limited policy learning. The learning of individuals, institutions and regions is an essential part of policy implementation.

International Engagement

Big science

A specific form of international engagement is active participation in large experimental facilities such as the European Organization for Nuclear Research (CERN) and the International Thermonuclear Experimental Reactor (ITER). From the innovation policy perspective, the interest is in commercializing the knowledge that results from pushing technology to its limit. An example is the contribution to medical imaging made by work on elementary particle detectors. There is also the World Wide Web which came out of CERN and which provides a platform for many unexpected commercial applications. Scientific organizations also provide postdoctoral training and develop the very highly qualified workforce.

International cooperation and development

Scientific cooperation among member states is an objective of the European Union, and Germany and Japan cooperate with developing countries as a way of addressing global challenges. There are scientific benefits, but the cooperation also builds knowledge of different markets and opens opportunities for commercial activity.

Germany and Japan are collaborating with developing countries as part of their approach to innovation policy. In the case of Germany, cooperation will support collaboration with research groups and innovative industry clusters with German research groups and competence networks. It is also supporting the Heiligendamm-L'Aquila Process (HAP) involving dialogue between the outreach group from the key emerging economies, the O5 (Brazil, China, India, Mexico and South Africa) and the G8 to address the promotion and protection of innovation and ways to increase energy efficiency (BMBF 2008b: 29). Japan's strategic promotion of Science and Technology (S&T) Diplomacy is designed to strengthen S&T cooperation with developing countries as part of resolving global issues, using Japan's advanced S&T. The global issues include the environment, energy, natural disaster prevention, infectious disease control and food security.

An OECD and United Nations Educational, Scientific and Cultural Organization (UNESCO) workshop on innovation for development in January 2009 concluded that innovation should be inserted in the Poverty Reduction Strategy Papers.[3] It also stressed the need for more knowledge about innovation in developing countries that could be produced through case studies or country reviews of innovation policy in developing countries, similar to those conducted by the OECD. The findings of the workshop are summarized in UNESCO (2009).

An incentive for including work with developing countries as part of an innovation strategy is the reduction of the inequities which are potential causes of conflict and disease, which can spread rapidly, and starvation. Innovation through collaboration can also foster a culture of innovation in the developing countries leading to economic growth and related benefits. Collier (2007: 11) points out that growth, as an objective, is not universally accepted in the development community unless it is qualified with terms like 'sustainable' or 'pro-poor', but argues that: 'the problem of the bottom billion has not been that they have had the wrong *type* of growth, it is that they have not had *any* growth'. Innovation strategies in developed countries and cooperation agreements have a role to play.

In April 2009, there was an OECD workshop on 'Innovating out of Poverty' which stressed the importance of recognizing agriculture as a knowledge-intensive sector, and the key role for science, technology and innovation contributing to this. The Chair of the workshop, Calestous Juma, provided a list of challenges for world leaders which is being disseminated (OECD 2009h).

Global Challenges

There are challenges that affect all countries, including developing countries, which can be addressed through innovation. These include climate change, sustainable energy, food and water security, and population health, as the world deals with the H1N1 pandemic in 2009 and prepares for the next. Green innovation (OECD 2009i) ensures that green activities are part of innovation, innovation policy and human resource development, and that they are part of the price signal.

SUMMARY

This chapter presupposes policy objectives of government which are going to be served by implementing the components that have been presented. The components offered are not meant to be exhaustive, but indicative of what could constitute a cross-cutting innovation strategy, to be implemented, monitored, evaluated and revised as part of the learning process. If any component is to be seen as essential, it is monitoring and evaluation, as without it being built in from the start, the learning opportunities are limited. In a rapidly changing world, with unpredictable dynamics, learning is part of survival.

In the next chapter, the issue is how to bring all or some of these topics together under an innovation strategy umbrella and to ask if this is possible or even desirable. To facilitate the discussion, a list of components is provided in Box 7.1.

BOX 7.1   POSSIBLE COMPONENTS OF AN INNOVATION STRATEGY

Component Activities

1. Markets
    1.1 Brand recognition
    1.2 Lead market
    1.3 Competitive engagement
    1.4 Financial services

2. People
    2.1 Labour force
    2.2 Demographics and demand for innovation
    2.3 Migration

3. Innovation Activities
    3.1 Technology and practices
    3.2 User innovation
    3.3 User-driven innovation
    3.3 Open innovation
    3.4 Demand-driven innovation

4. Public institutions
    4.1 Infrastructure
    4.2 Procurement
    4.3 Priority setting
    4.4 Standard setting
    4.5 Public finance
    4.6 Government direct support
    4.7 Education, training and research
    4.8 Health
    4.9 Monitoring and evaluation

5. International engagement
    5.1 Big science
    5.2 International cooperation and development
    5.3 Global challenges

NOTES

1. Further information on the UK SBRI is available at www.innovateuk.org/deliveringin-novation/smallbusinessresearchinitiative.ashx.
2. Information on NRC-IRAP is available at www.nrc-cnrc.gc.ca/eng/ibp/irap/about/index.html.
3. The Poverty Reduction Papers are described at www.imf.org/external/NP/prsp/prsp.asp.

8. Innovation strategy coordination

INTRODUCTION

Innovation strategies consist of a set of components that are managed, or coordinated, by government as part of their implementation. They range from a single intervention managed by one department, such as a tax credit for expenditure on information and communication technologies (ICTs), to a wide range of initiatives drawn from the components in the previous chapter and requiring a 'whole of government' approach. There are also cases of more than one innovation strategy being managed at the same time by a government, with one perhaps led by the Finance Department, one by the Research or Industry Department, and one by the Education or Human Resources Department.

As the number of components grows, so does the challenge of coordination. Depending upon the government and the management culture, it can be very high level, such as Cabinet chaired by the head of government, a Cabinet committee chaired by a senior minister, a committee or council of stakeholders in the economy and society providing advice to a senior minister, or lower levels of coordination managed by interdepartmental working groups or even by departmental working groups. The difficulty with innovation policy is that it crosses many departmental boundaries. In some countries, research and development (R&D) and capital investment incentives are tax matters; direct grants, contracts and contributions cross all departments and they include procurement policy.

As all countries differ in their history, culture and innovation system, there is no single answer to how many components there should be in an innovation strategy and how those components should be coordinated and at what level. This chapter looks at the experience of a number of countries and draws inferences but not definitive conclusions.

COMPONENTS AND COORDINATION

Levels of Coordination

In Chapter 7 a list of possible components of an innovation strategy was produced. The question to address in this chapter is how to coordinate the activity. This is motivated, in part, by the reasons for introducing an innovation strategy.

If the leaders of the country believe that the global challenges discussed in Chapter 1 have to be addressed, as well as the domestic issues, such as competing in a global economy with an ageing and diminishing labour force, the strategy will be coordinated at the highest level. In a parliamentary government, the lead would be the Prime Minister and the committee would be the Cabinet, or a selection of Cabinet ministers.

If the concern is with industry, and its ability to compete, the coordination could be given to the minister responsible for industry. Similarly, if the issue is the need for a better-educated and trained population engaged in lifelong learning, the coordination could be left to the Minister of Education. If the issues are seen to be sectoral, such as helping the service sector to be more competitive, the coordination could be done at sub-department level within a ministry.

All of this is a process happening under a single administration, and it is a major undertaking of policy design and implementation, but continuity of innovation strategy from administration to administration is another issue. Solving this is beyond the scope of this book but the hope is that good policies, well coordinated, monitored, and shown to deliver results through evaluation, would be difficult to discontinue in the event of a change of government.[1]

Engaging the Stakeholders

Innovation is about bringing products to market. Innovation strategies have little hope of succeeding if industry leaders are not part of the discussion. As people are key to all strategies, leaders in education and training should be involved, as well as the government leaders promoting the strategy. Then, there is civil society, which will include the consumers responsible for user-driven innovation and those who are user-innovators. As innovation has both social and economic impacts, there is a case for the participation of industry associations and representatives of organized labour.

This suggests that in addition to well-thought-out coordination there should be a role for a council of stakeholders to contribute to the formulation of the strategy and to provide advice on its implementation. Such a council is not a science or research council, or a science and technology council. It should be an innovation council, as the issue is developing new products and processes to compete locally and globally and to give rise to sustainable growth through improved productivity. This is a very concrete goal which is more immediate, given the recession of 2008–09, than the important but longer-term issue of supporting the formal generation of knowledge through the performance of R&D. Most industrial innovators do not do R&D.

The coordination, and the components coordinated, will reflect the governance structure, history and culture. Federal governments will act differently from central governments, as will multicultural countries from those that are more homogeneous. That is why there is no single innovation strategy. In what follows, some country experience is examined to see what was being done in 2009.

COUNTRY PRACTICES

The 2008 *OECD Science, Technology and Industry Outlook*, OECD (2008c), provides a two-page Science and Innovation Country Note for OECD countries: one page of narrative and one of indicators. In this section, examples will be taken from a selection of countries to illustrate the diversity of approaches to innovation strategies. The selection is not intended to be exhaustive and only a few points are made for each country. Readers are encouraged to look at the Country Notes in OECD (2008c) and to read the country documents.

In addition to the Country Notes, there are the OECD reviews of innovation policy conducted at the invitation of member and observer countries.[2] An example is the review of the innovation policy of Norway (OECD 2008c). These provide in-depth analyses of innovation policies currently in place, and there are findings of earlier reports summarized in OECD (2005e). Country reports are done by other organizations such as the Observatoire des Sciences et des Techniques in Paris, the United Nations Educational, Scientific and Cultural Organization (UNESCO) and the World Bank.

Denmark

The Danish *Strategy for Denmark in the Global Economy* was published in 2006 (Government of Denmark 2006) and focuses on education and learning, research, interaction with other countries and cultures, and support for high-growth start-ups. It includes a Globalization Council which brings together a wide range of stakeholders and government officials and provides government oversight.

Finland

In 2008 the Government of Finland presented a Communication on Finland's National Innovation Strategy to the Parliament based (Government of Finland 2008a) on an earlier Proposal for Finland's National Innovation Strategy (Government of Finland 2008b). The Innovation Strategy provides a comprehensive example of the use of strategic components and their coordination.

The strategic goals of the strategy are led by growth, combined with the well-being of people and the environment. This gives the environment a high level of visibility. Public sector innovation is seen as one means of achieving the goals, along with private sector innovation-led productivity growth.

The second goal is pioneering in innovation activity which stresses that innovation policy must cross administrative boundaries, support technological and non-technological innovation, and encourage a culture of innovation. The need for Finns to influence the goals of regional, national and international development is part of the strategic choices.

The proposal for Finland's National Innovation Strategy covers most of the points in the generic list, but the real content is in how it puts these points together. Brand recognition is important for the attraction of people and investments based on the strategic choices which exemplify Finnish priority-setting.

There is emphasis on demand-driven innovation and involving the users and clients in the innovative process. This is not user-driven innovation as defined in Chapter 1, involving a collaborative interaction between users and producers. This user–producer network is part of the networks that characterize Finish institutions and it recognizes that innovation policy should bring together the needs of users, consumers and citizens alongside efforts to build knowledge, creativity and competence.

The success of the European Union (EU) economic and innovation policy is important to Finland but it is also important for Finland to engage and influence the EU as well as taking advantage of all of the instruments of EU innovation policy. Four drivers of change are singled out: globalization, sustainable development, new technologies and the ageing population of Finland. The last point raises this human resource issue to a higher level than that found in some other strategies.

The strategic choices, or priorities, in the Strategy start with global engagement including the already mentioned need to engage at all levels of decision-making, as well as the need for Finns to be mobile and for the country to be attractive to people and as a place to invest. The second choice is emphasis on demand for innovation and the link between producers and users. Encouraging individuals, innovation communities and entrepreneurs is the third choice, and a systemic approach including a determined management of change is the fourth.

Throughout the Strategy, implicitly or explicitly, is the importance of networks and participation in the networks as a priority. Not just international and national networks, but also those which connect to regional centres of innovation excellence. The networks also illustrate the inclusiveness of the Strategy and its recognition of the arts and nature as sources of experience and new ideas. These new ideas contribute to electronic content, which is a growing area of economic activity.

The Strategy includes ten actions to be taken and an implementation plan. The breadth of the Strategy ensures that it will be managed across government and beyond, and there is a history in Finland of long-term investment to support orderly development over time. The Cabinet Committee on Economic and Innovation Policy will manage the implementation of the Strategy and a Research and Innovation Council was created in January 2009, chaired by the Prime Minister, with senior cabinet ministers as members as well as stakeholders from business and civil society.

In the Country Notes (OECD 2008d: 116), there is an observation that the high R&D intensity of Finland has yet to be converted to the expected innovations, jobs and exports; there are few R&D spin-offs because of a lack of venture capital; and there is little co-patenting with foreign co-inventors. The apparent problem with the connections needed for commercialization contrasts with the importance of network participation in the proposed strategy. However, the focus on influencing supranational and international organizations is key to integrating Finland in the world economy. Networks matter, and the network capital accumulated in them is critical.

The role of knowledge in the Finnish innovation system has previously been examined in the course of a review by the World Bank Institute (Dahlman et al. 2006). Von Hippel would argue that the strategy could be strengthened by the addition of user innovation, whether the users are firms or individuals.

France

France is embarking on the development of an innovation strategy, after review by the Haut Conseil de la Science et de la Technologie (HCST) and presentation to a ministerial council. It takes a cross-cutting approach and is planned to address four families of challenges: social, including the ageing population and global food and water supply; knowledge, including the areas where France should engage either on its own or in collaboration (mathematics, social sciences and humanities, life sciences, physics and the work at the European Organization for Nuclear Research (CERN) and the International Thermonuclear Experimental Reactor (ITER) are examples); the need to master key technologies (bio- and nano-technology, ICTs, and technologies for sustainable development); and organizational, including knowledge flows, the integration with EU policies, and with the European Strategy Forum on Research Infrastructures (ESFRI).

The approach involves contributions from ministries, coordinated by the Ministry of Higher Education and Research. In addition, wide involvement of stakeholders is planned through the Internet and working groups (République Française: Ministère de l'enseignement supérieur et de la recherche 2008). There is also a review of the approaches to developing innovation strategies in Denmark, Finland, Germany, Japan and the UK, and how they align with that of France.

Germany

German policy related to innovation is found in *The High-Tech Strategy for Germany* (BMBF 2006) and in *Strengthening Germany's Role in the Global Knowledge Society* (BMBF 2008b). The high-tech strategy is an example of priority-setting and it sets out 17 technologies in three broad categories: innovation for a safe and healthy life; innovation for communication and mobility; and innovation through cross-cutting technologies. The strategy promotes: the exchange of knowledge, and experts embodying the knowledge, between research institutes and industry; improving conditions for high-tech start-ups and innovative small and medium-sized enterprises (SMEs); supporting rapid diffusion of new technologies; and strengthening Germany's international position. In doing this, the importance of a coordinated innovation policy is stressed, along with the potential for Germany to become a lead market, while playing an active and competitive international role.

The strategy of the federal government for the internationalization of science and research (BMBF 2008b) has four main goals: strengthening research and cooperation with global leaders; international exploitation of innovation potentials; intensifying the cooperation with developing countries in education, research and development on a long-term basis; and assuming international responsibility and mastering global challenges.

The third point has similarities with the Japanese Science and Technology Diplomacy initiative (Council of Science and Technology, Japan 2008).

A more recent document (BMBF 2008c), *10 Thesen für ein starkes Wissenschaftssystem im weltweiten Wettbewerb, Demands on Research Landscapes under Changing Framework Conditions,* provides the outcome of two rounds of expert consultations, summarized in ten theses which deal mainly with research excellence, competition for funding, research institutions, the European Research Area, mobility of the labour force and the transfer of knowledge to industry. It is this last item that raises a different view on the technology transfer process advocated in many countries for moving knowledge from the universities or research institutions to the private sector for commercialization.

The experts saw professional technology transfer as a means of contributing to the brand name of the institution, enhancing visibility, attracting new faculty members, building contacts with the private sector, facilitating fund raising and making the political decision-makers more aware of the institution's role in advancing economic and social welfare. They recognized that it takes 15 to 20 years of professional technology transfer before there are even modest financial returns. This is a quite different, and broader, perspective from that found in some research or innovation policies which focus on the commercial value of technology transfer.

The high-tech strategy and the internationalization strategy, with the expert comments on policy in BMBF (2008b), provide a view of the German approach to an innovation strategy. It is overseen by a Council for Innovation and Growth, which advises the Chancellor and is supported by the Industry–Science Research Alliance on the Technology Prospects of Markets of the Future, created by the Federal Ministry of Education and Research (BMBF) and involving representatives from the industry and science sectors. There is also provision in the implementation plan for regular evaluation of progress.

The Netherlands

The Netherlands has an agenda for sustainable growth in productivity (Government of the Netherlands 2008). It stresses the importance of strengthening and exploiting talent, and knowledge from public and private research, and the promoting of innovative enterprise. Target areas are water, building upon Dutch expertise, logistics, fuels from non-edible sources, food and nutrition, and security. The Netherlands has also been experimenting with an Innovation Voucher subsidy scheme (SenterNovem 2006) as part of promoting knowledge flow in the innovation system. The agenda sets out targets and examines how it will reach the goals set for

2030. In setting goals beyond the mandate of the present government, the policy deliberately takes a long-term view.

The agenda is based, in part, on work of the Innovation Platform (2009), which was created in 2003 to stimulate, but not fund, innovation in the Netherlands. It is chaired by the Prime Minister, has senior ministers as members, and other members from public and private institutions. It provides an oversight function for innovation initiatives. The agenda is part of a government programme entitled 'The Netherlands: Land of Enterprise and Innovation', which includes the drafting of social innovation agendas in the areas of security, water, energy and healthcare. The approach to innovation policy in the Netherlands is comprehensive, with targets, and with a high level of oversight.

Sweden

Sweden published its innovation strategy in 2004 (Government of Sweden, Ministry of Industry, Employment and Communications 2004) well before the European Innovation Strategy. It presents globalization as an opportunity, and promotes education, entrepreneurship, enterprise and skill development, the innovative capacity of SMEs, and commercialization of research and ideas. In the public sector the focus is on renewal efficiency and sustainable development. It highlights the advantage of being a diverse population with a significant portion of the population having been born abroad.

The importance of coordination and collaboration between policy areas was seen as a key task in implementing the strategy and governance of innovation systems, with Sweden as a case study, is discussed in OECD (2005e). The OECD (2008c) country report notes the need to increase innovation in service industries in Sweden.

The United Kingdom

In the UK, the Sainsbury report (Sainsbury 2007) and the 2008 Enterprise Strategy (HM Treasury/BERR 2008) influenced the 2008 innovation White Paper (DIUS 2008) which is a comprehensive policy document stressing the role of design in innovation and the need to know more about intangible investments. Procurement is seen as an opportunity for innovation and service industries, and creative industries are singled out along with open innovation and better regulation. The White Paper is to be followed with an international strategy and a science and society strategy.

In parallel with the White Paper and the supporting reports, the UK Department for International Development (DFID) produced a research strategy for 2008–13 (DFID 2008) which commits to striking a balance between knowledge and technology creation and getting new and existing technology into use.

In 2009 a paper, *New Industry New Jobs* (HM Government 2009), presented an activist industrial policy to deal with the economic downturn. Its components could form part of an innovation strategy. The Technology Strategy Board, along with regional bodies, has been promoting innovation vouchers to allow SMEs to purchase knowledge from universities and colleges, and there are schemes to support commercialization of new ideas from universities. There has also been the establishment of the Small Business Research Initiative (SBRI) to link procurement and SMEs.

In June 2009, the Department for Innovation, Universities and Skills (DIUS) and the Department for Business, Enterprise and Regulatory Reform (BERR) merged to become the Department for Business, Innovation and Skills (BIS).

Asia and the Americas

The Organisation for Economic Co-operation and Development (OECD) countries in these regions are Australia, Canada, Japan, Korea, Mexico, New Zealand and the United States. Canada, Japan and the United States are selected as examples.

Canada

The Canadian science and technology strategy, *Mobilizing Science and Technology to Canada's Advantage* (Government of Canada 2007) was released in 2007 by the Department of Industry and the Department of Finance to signal both its importance to government and the cross-cutting nature of the policy. The policy is focused on science and technology. It deals with business (entrepreneurial advantage), research and development (knowledge advantage) and people (people advantage) and the policies relevant to promoting science and technology (S&T) in each domain. In doing this it recognizes the importance of S&T to business, especially in view of the lower productivity of Canada compared with that of the US (CCA 2009a, 2009b). It also recognizes the need to strengthen the knowledge base and to be able to attract people to work in Canada. This last point acknowledges the problem of the ageing population and the ability of Canadians to leave the country to work anywhere in the world.

While strengthening the knowledge base is a goal, it is tempered by limitations on resources. As a result, there are four priority areas: environmental science and technologies; natural resources and energy; health and related life sciences and technologies; and information and communication technologies (ICTs). These acknowledge the history of the country, with strong resource and energy sectors and corresponding academic engagement; the need, in a large country, for an ICT infrastructure; the demands of an ageing population on healthcare; and the environmental impacts of the resource and energy sectors.

The strategy commits to creating a climate of innovation and discovery by providing an enabling environment for business, supporting the production of the next generation of researchers, and being accountable for delivering results which will ultimately benefit Canadians. Targets are not part of the strategy but directions, and progress in these directions, can be monitored and the means of making the progress evaluated. From the innovation strategy perspective, this is an enabling policy.

There are 36 policy commitments in the 2007 strategy, one of which is to consolidate a number of sources of policy advice and to create the Science, Technology and Innovation Council (STIC), reporting to the Minister of Industry, and charged with providing policy advice to the government on S&T issues and with producing regular State of the Nation reports that benchmark Canada's performance against international standards of excellence. The first report (Government of Canada 2009) of the new Council examined the state of the science, technology and innovation system, with an emphasis on innovation.

The final policy commitment is to increase the accountability of the federal government by improving its ability to measure and report on the impact of S&T expenditures. While this is consistent with recommendations of the OECD Blue Sky II Forum of 2006, impact measures remain a challenge.

From the coordination perspective, the S&T strategy is managed by a single department, the Department of Industry, and the STIC advises the Minister of Industry, although it draws from people in business and public institutions. In Canada there is no federal department of education and the Department of Human Resources and Skills Development is not implicated in the policy commitments.

Japan

The place of innovation in Japanese policy is well summarized by Harayama (2007) who reviews the Science and Technology Basic Law, 1995, and the Basic Plans that followed it: the First, 1996–2000; the Second, 2001–05; and the Third, 2006–10. It is in the Third Basic Plan that innovation appears, but much that would contribute to an innovation strategy was introduced under previous plans. Under the Second Basic Plan a Council for Science and Technology Policy was created and priorities were set: life sciences; information technology; environment; nanotechnology; materials; energy and manufacturing technology; social infrastructure; and new frontiers likely to bring solutions to major problems.

Under the current Third Basic Plan, there are plans to accelerate innovation by building Centres of Excellence, stimulating interdisciplinary fields, and developing human resources. Part of the human resource initiative is to encourage mobility and to attract foreign researchers.

Harayama (2007) also reviews the 'Innovation 25' initiative which sets the target of making Japan one of the most innovative countries by 2025. From the perspective of international cooperation, she presents Science and Technology Diplomacy and plans to address environmental problems through international cooperation. There are similarities with the German approach (BMBF 2008b: 27).

Compared with the US, the Japanese Third Basic Plan, Innovation 25, and Science and Technology Diplomacy provide a more coherent and centrally managed approach to the promotion of innovation, one that is closer to that of the European countries.

The United States

Historically, the US has not taken a whole-of-government approach to innovation. Until 21 September 2009 there was no document which could be regarded as an innovation strategy and there have, until the introduction of the Business R&D and Innovation Survey in 2009, been no ongoing surveys of the activity of innovation in US industry along the lines of the Community Innovation Surveys in Europe (Parven 2007). This does not mean that there has not been interest in innovation or that there are no policies that promote innovation.

A National Research Council Symposium, 'Innovation Policies for the 21st Century', Wessner (2007), looked at innovation policies in other countries and their relevance for the US. It took note of the US study *Rising Above the Gathering Storm* (National Academy of Sciences/ National Academy of Engineering/Institute of Medicine 2007). Another National Research Council (NRC) study (Macher and Mowery 2008) presented ten industry studies from the perspective of innovation in global industries, following a previous and much cited study of competitive performance in selected industries (Mowery 1999) which also influenced the work reported by Wessner (2007). Atkinson (2004) reviewed the US economy and made recommendations that could be part of an innovation strategy. Jaffe et al. (2006) looked at innovation policy from the intellectual property perspective. There are many scholars, and officials, in the

US working on aspects of innovation policy and strategies and producing reports which have influence across government, but not in a coherent or coordinated manner.

Measurement of innovation has also been a topic of ongoing discussion. A National Research Council panel reviewed the measurement of R&D and innovation at the National Science Foundation (NSF) (Brown et al. 2005). An Advisory Committee to the Department of Commerce issued a series of recommendations on ways to improve the measurement of innovation in the US economy (US Department of Commerce 2008). The Conference Board ran a workshop on 'Developing a New National Research Data Infrastructure for the Study of Organizations and Innovation' in July 2008 (Conference Board 2008) and there was a *Business Week* article on the NSF initiative in September 2008 (*Business Week* 2008).

Policy issues are being discussed and measurement of innovation is being introduced. At the same time the NSF is supporting research on the Science of Science and Innovation Policy (SciSIP) to improve the understanding of innovation policy, to broaden the discourse, to improve existing data, and to build new benchmark data series to facilitate the study of innovation. An important objective of the SciSIP programme is to build communities of practice among scholars (Chapter 10). This is in addition to the procurement programmes of Defense, Energy, Homeland Security, and other parts of the US government, and the work of the Small Business Innovation Research (SBIR) programme which was recently reviewed (Wessner 2008). These are not whole-of-government coordinated activities, but there are many activities supporting debate on, research into, measurement and evaluation of, and support for innovation. This is a different model from that of European countries and Japan.

The approach to innovation strategy in the US does raise the question about the size of the country and the economic impacts of its policies. Any policy initiative of the Department of Energy, for example, will have large impact. The question to resolve is whether it is better to manage that initiative well, or to coordinate with other initiatives that might increase transaction costs without providing a significant return on the investment in coordination.

Meanwhile, President Obama, in an address to the National Academy of Sciences (Obama 2009), presented a number of commitments of his administration that would fit into an innovation strategy. The US will spend more than 3 per cent of GDP on R&D, which exceeds the Lisbon target of 3 per cent, but it is not allocated to the business sector (2 per cent) and the government sector (1 per cent). The R&D tax credit programme will be made permanent, allowing firms that benefit from it to make it part of their business strategy. In view of the energy crisis, an Advanced Research Projects Agency – Energy, ARPA-E, is being funded to undertake high-risk, high-return research. There is also a high-level council in the form of the President's Council of Advisors on Science and Technology (PCAST), which will be expanded. PCAST, however, provides S&T advice. There has been a suggestion (Atkinson and Wial 2008) for a National Innovation Foundation.

Following upon his announcement of an R&D target, President Obama released *A Strategy for American Innovation: Driving Towards Sustainable Growth and Quality Jobs* (Executive Office of the President 2009). This is a major initiative. It addresses the 'Grand Challenges' and the need for better health technology (to support an ageing population), supports advanced vehicle technology and unleashes a clean energy revolution. It covers the ground discussed earlier including the need for jobs and sustainable growth. As in the OECD and EU strategies, public sector innovation is an issue, as is community innovation.

The question for observers of the US Innovation Strategy is how it will be implemented, monitored, evaluated and adjusted to deliver the desired outcomes. That leads naturally to a discussion of coordination practices.

COORDINATION PRACTICES

Of the examples presented, all but the US and Canada attempt a whole-of-government approach to innovation strategies and most involve stakeholders from outside of government. Human resource issues are prominent and the ageing population is a recurring motivation to address the resulting issues through innovation. Most strategies recognize the need for the skilled population to be mobile and they identify the need to retain such people in order to create knowledge and to be competitive globally. Given the demand, there is a question as to whether there are enough of the highly qualified to go around. Some countries, explicitly Germany and Japan, are looking to collaborate with developing countries as part of their innovation strategies. Framework conditions are pervasive, with focus on regulation, procurement and the importance of supporting lead markets, while acting and trading competitively abroad.

While the European countries and Japan take a more coherent approach, there are many mechanisms in the US supporting innovation. Given the size of the US market, any one of the policies there, such as procurement for the Department of Defense, could have significant economic and social impact which could be reduced by integrating it with other policy initiatives. This remains an open question. However, it is about to be answered as the President's Innovation Strategy is implemented.

SUMMARY

This chapter has examined how some governments have chosen to integrate some of the policy components discussed in the last chapter in order to have an innovation strategy. There is a difference in approach between Europe, in which countries attempt a whole-of-government approach, and North America where there is less coherence. The effectiveness of the different approaches is a subject for greater study as innovation policies evolve, but the size of the economy in which the strategy is implemented is clearly an issue.

The need for monitoring and evaluation are recurring themes in this book because these activities lead to policy learning and better policies as a result. With so much activity around innovation strategies and their implementations, there is an opportunity to do comparative analysis at the country level of successes and failures, and to model the global interactions that affect these outcomes. The knowledge gained about the global, complex, dynamic and non-linear system could be used to help those countries that have a long way to go to gain significant benefits from innovation. That is the subject of the next chapter.

NOTES

1. That this is a very optimistic hope is recognized. The reader may wish to find examples of policy discontinuities and continuities which could become a 'science of innovation policy' paper.
2. A list of OECD innovation policy reviews may be found at www.oecd.org/sti/innovation/reviews.

PART IV

Extending the community and the subject

9. Innovation and development

INTRODUCTION

This chapter looks at innovation measurement and policy in developing countries and the extent to which these activities differ from those in the developed economies. While there are differences of emphasis, the basic definitions given in Chapter 3 still apply. What is needed is the accumulation of experience of measuring innovation, interpreting the results, and using the findings in policy development, monitoring and evaluation. The challenge is implementing, in the context of development, what has been learned over the last 30 years.

Studies in developing economies suggest that innovation is more incremental than radical, but that is just a difference of degree from the developed economies. In the developing world, the informal economy plays a larger part and, by its nature, it is not accessible to standard survey methods. It is an area for case studies, as all of the components of innovation are present, the producers, the suppliers and the market; the innovation can be studied by using structured interviews. The results may highlight the need, for example, to treat agriculture as a knowledge-based industry in a global world, rather than a subsistence activity, or the need to protect indigenous knowledge so that its use can continue to benefit the community that has developed it over time.

A strong business sector is a characteristic of a developed economy, but not necessarily of a developing one. Those firms that are present may focus more on innovation for survival than on formal knowledge creation through research and development (R&D) activities, and they may have a low capacity to absorb knowledge from outside of the firm needed to create value and put new products on the market. While the business sector may be small in developing countries, the agriculture sector can be quite large which makes it a fruitful domain for innovation policies and their application. With growing urbanization, manufacturing and service firms have a greater opportunity to participate in innovation clusters.

Support for innovation is more of an issue in developing countries. The infrastructure, such as broadband Internet access, water and reliable electricity supply, roads, ports and basic telecommunications services may not be sufficiently well established to facilitate business activities. Framework conditions, such as courts, education, stable governance, health services, security and tax systems may not align to support innovation in the private sector. As innovation is an outcome of an innovation system, system alignment or misalignment, or system failures, are issues for consideration.

Innovation is linked to growth in industrialized countries, although the financial crisis of 2008–09 is linked to innovation in financial services. The link between innovation and growth is less evident in developing countries (RICYT/OECD/CYTED 2001), in part because of the size of the business sector, but that does not make innovation any less of a topic of policy interest, and a challenge for measurement (Blankley et al. 2006).

Monitoring and evaluation are part of policy learning that results from case studies findings, statistical indicators, and analysis of the outcomes and impacts of policy. There are also international policies to monitor and to evaluate, such as those embodied in the Millennium Development Goals (MDGs).[1] Innovation and innovation policy have roles to play in moving towards the goals in 2015.

The chapter is not a comprehensive guide to innovation measurement and policy in developing countries. It is an illustration, with selective examples, of similarities in approaches with those used in developed countries. However, the result need not be the same as in developed countries as different priorities may suggest a different selection of components of an innovation strategy and the coordination mechanism, or mechanisms, would have to fit with the existing culture and governance practices.

For those who wish to read more broadly, Aubert (2004, 2006) has proposed a conceptual framework for promoting innovation in developing countries and Scerri (2006a, 2006b) examines contexts for measurement in developed and developing countries. Rodrik (2007) deals with different paths to economic growth and Collier (2007) addresses the problems of people in the poorest countries, the 'bottom billion'. The *Industrial Development Report 2009* (UNIDO 2009) provides an overview of the industrial challenges for the bottom billion and for the middle-income countries, and innovation plays a part in both. A chapter in the World Bank *Global Economic Prospects* for 2008 deals with measuring technological diffusion in developing countries (World Bank 2008: 51). As mentioned in Chapter 1, both Diamond (1997) and Wright (2004) have insights on innovation and technological change in development situations and illustrations of what has worked and what has not.

INDICATORS AND LANGUAGE

In comparison with developed countries, innovation in developing countries takes place in small firms or in the informal economy, and with limited support from infrastructure and framework conditions. Firms, which may be individual entrepreneurs, often lack the absorptive capacity to seek knowledge and to absorb it into the firm so that it can be converted to value as part of the innovation process. Firms do work with existing technologies and adapt them to their needs, and they are able to organize their labour force and use of management practices in order to become more productive.

All of these topics have been covered in earlier chapters, and how they are addressed influences measurement. Examples from Latin America and from Africa provide examples of two approaches to measuring innovation in developing countries.

Latin America

In Chapter 4, there was a discussion of the use of the Oslo Manual as a guide for measuring innovation and how some users were selective in their use and others took the view that the manual did not meet their needs and then developed their own. An example of a community developing its own standards is provided by the Bogotá Manual (RICYT/OEC/CYTED 2001) in Latin America. The Bogotá process brought together experts in Latin America over a number of years and gave rise to discussion and consensus building around the problem of measuring innovation in such a way that comparisons could be made in Latin American countries, using the resulting indicators (Lugones 2006). It built upon the experience gained from conducting surveys of innovation in some Latin American countries in the 1990s (Anlló 2006).

The issues addressed in the Bogotá Manual were those discussed in Chapter 4 for firms that were innovative but did not conduct R&D. Innovation resulted from the investment in machinery and equipment and in change to the organization of the firm or the use of new business practices. For these activities to happen, the firms required access to sources of information for innovation from their clients, suppliers or competitors. A list of sources used in the EU Community Innovation Survey is given in Chapter 4 and Chapter 7 of the Bogotá Manual.

An important element of the Bogotá Manual is its focus on human resource issues and how they are organized, on training and human capital, and on participation in networks. Also addressed are the acquisition of technology and knowledge, and the likelihood of the adaptation of the technologies acquired. This is not extensively developed but it links to the work on user innovation defined in Chapter 1 and discussed in Chapter 4. The use of human resources, technologies and external knowledge are key characteristics of innovation in all countries and discussions of these topics were also going on, at the same time, in the Organisation for Economic Co-operation and Development (OECD) and European Union (EU) communities as well as in Latin America.

In the first chapter of the Bogotá Manual there is reference to the complementary consequences of conducting surveys as part of developing indicators:

> The surveyed or interviewed firms and/or institutions are, in the first place, forced to reflect on their actions in the fields of science and technology and, secondly, subsequent discussion of the information gathered makes it possible to discover associations and links between performance and any action taken in connection with technological change. (RICYT/OEC/CYTED 2001: 10)

The first consequence relates to the survey as a teaching instrument as well as a means of information gathering, and the potential use of surveying as a policy tool. The second point deals with the understanding of causality through interviews and discussion. This is an important point as causal relationships cannot be inferred from observations from repeated cross-sectional surveys or case studies (Chapter 5).

The Bogotá Manual was a milestone in indicator development, and not just in Latin America (Gault 2008a). It contained many of the elements that would later be elaborated in the third edition of the Oslo Manual (OECD/Eurostat 2005) during the discussions of 2003–05. The Bogotá Manual grew out of the work of a scholarly community in Latin America supported by the Network on Science and Technology Indicators Ibero-American and Inter-American (RICYT), the Organization of American States (OAS) and the Latin-American Science and Technology Development Programme (CYTED).

As the Manual emerged from a special project conducted over a number of years, it did not have the ongoing support of a professional secretariat to manage its implementation, or revision, or a permanent body of government officials, such as the OECD Working Party of National Experts on Science and Technology Indicators (NESTI), to provide feedback on the country experience of running innovation surveys based on the Manual. The considerable cost of maintaining a manual which evolves over time may have been a factor in an initiative taken by RICYT in 2004.

In 2004 RICYT proposed to NESTI (RICYT 2004) that the third edition of the Oslo Manual, which was then being discussed, include an Annex interpreting the Manual for use in developing countries. This was agreed by the 2004 NESTI meeting and the UNESCO Institute of Statistics (UIS) agreed to coordinate comments on the Annex from developed and developing countries and to produce a draft of the Annex. The Annex was approved by NESTI in 2005 and included in the third edition.

The advantage of this important act was that the language of discourse was expanded, rather than fragmented, and developing countries, or emerging economies with different measurement challenges from those of OECD countries, could look at ideas for future guidelines in three ways. The first was consideration of topics that were specific to the economic situation of the countries and could give rise to interpretive texts for national or regional use. Next was the review of topics that raised questions fundamental to the future of the Oslo Manual, which would contribute to the discussion leading to the next revision. And finally, the review of topics that were somewhere in between, discussion of which could contribute to a revised Annex.

This approach is also applicable in developed economies as every country is different and, in the course of measuring the activity of innovation, questions arise that have to be resolved in the context of the country's economy and society. In principle, every country could have a document providing guidelines on the application of the Oslo Manual.

The RICYT initiative concerning the Oslo Manual gave rise to a proposal for a similar Annex to the Frascati Manual (OECD 2002b) and as of 2009 the input to that Annex is also being coordinated by the UNESCO Institute of Statistics. There is no reason why similar annexes could not appear in other members of the Frascati Family of manuals (OECD 2002b: 16).

Africa

The approach to indicator development in Africa differs from that in Latin America as it evolved as part of a high-level initiative of the New Partnership for Africa's Development (NEPAD) and the African Union (AU). The African Union was formed in 2001 with 53 member states and succeeded the Organization of African Unity (OAU) and the African Economic Community (AEC). NEPAD is a programme of the AU.

In 2003, the First NEPAD Ministerial Council on Science and Technology took place in Johannesburg, South Africa, and resolved to 'Develop and adopt common sets of indicators to benchmark our national and regional systems of innovation' (NEPAD 2003). The Council also adopted the outline of a plan of action as the basis for the formulation of NEPAD's business plan on science and technology. What became the Consolidated Plan of Action (CPA) (NEPAD 2006b) was adopted at the second African

Ministerial Council on Science and Technology[2] (AMCOST II) (NEPAD 2005a) and the same meeting agreed 'to establish an intergovernmental committee or relevant national authorities to develop, adopt and use common indicators to survey and prepare an African Science, Technology and Innovation report'.

Leading up to AMCOST II, and the decision to establish an intergovernmental committee, a group of experts met to consider how to proceed towards developing African indicator manuals and the possibility of establishing an Africa observatory for the collection of science, technology and innovation (STI) data, the analysis of the data and the development of indicators (NEPAD 2005b). The experts, with the NEPAD Secretariat, also prepared terms of reference (NEPAD 2005c) for the intergovernmental committee.

In 2006, the Sixth Meeting of the Steering Committee of the African Ministerial Council on Science and Technology (AMCOST) (NEPAD 2006c) reviewed the terms of reference for the Intergovernmental Committee on Science, Technology and Innovation Indicators and provided very clear direction as to what was expected. It resolved that: 'ways and means should be explored to ensure that existing appropriate international indicators are adopted and used to survey science, technology and innovation in Africa'. An extraordinary conference of AMCOST followed (NEPAD 2006d).

The Steering Committee also called for joint efforts on the part of the NEPAD Office of Science and Technology and the AU to consult and to submit to AMCOST a comprehensive strategy for implementing the CPA. In Section 4, Programme 5.1 of the CPA, African Science, Technology and Innovation Indicators (ASTII) initiative, there was a project that dealt with indicator development (Kahn 2008), capacity building, international participation and the provision of information about the state of STI in African countries. A second project was the establishment of an Observatory. In the same year, a proposal was made to the Swedish Agency for Research Cooperation (SAREC) of the Swedish International Development Cooperation Agency (SIDA) (NEPAD 2006e) to enable African countries to develop common indicators, train government officials to conduct science, technology and innovation systems surveys and develop related strategies and policies, develop guidelines for policy review and development, establish a consortium of leading institutions to support countries to build policy analysis skills, and generate the first African STI Outlook. The proposal was successful.

In 2007, a number of things happened to move the indicators initiative forward. The OECD Working Party of National Experts on Science and Technology Indicators (NESTI), and its Secretariat, invited the NEPAD

Office of Science and Technology to participate in the 2007 NESTI meeting as an observer on the same basis as RICYT. This ensured that Africa had a voice at the table where international standards were set. The development of indicators began, supported by Sweden, and the first meeting of the African Intergovernmental Committee on Science, Technology and Innovation Indicators (NEPAD 2007) took place in Mozambique.

The Intergovernmental Committee, after thoughtful discussion, decided that:

> African countries shall use the existing internationally recognized STI manuals and/or guidelines, particularly the Organisation for Economic Co-operation and Development (OECD) Frascati and Oslo Manuals to undertake Research and Development (R&D) and innovation surveys respectively. They may use these manuals, and experience gained in undertaking surveys, to develop African STI manuals or guidelines.

This was a key decision which allowed the STI survey project to move forward rapidly (Gault 2008b).

In 2008 the NEPAD Office of Science and Technology, with SIDA support, held its first workshop on the use of the Frascati and Oslo Manuals in support of surveys in 19 African countries, and a second workshop was held in 2009 to discuss problems arising from the survey work either done or planned. It became evident that a community of experts was emerging with the capacity to support other measurement activities in the remaining 30 or so African countries in the next few years. These experts could establish the African equivalent of the OECD NESTI Working Party.

The surveys are expected to produce new information, as well as an expert community, and that information is to be used by the NEPAD Office of Science and Technology to develop and publish in 2010 the first *African Innovation Outlook* which will inform the people of Africa about STI activities in their countries. This is a major step in the development of indicators and analysis in support of evidence policy.

The only part of the CPA programme which has yet to be implemented (as of 2009) is the STI Observatory, on which discussion continues at the AU. If it is established, it would be the logical recipient of the aggregate data from the STI surveys and a centre of analysis and publication. The Observatory could also collect and review STI policies which are implanted in African countries in order to share best practices and improve the effectiveness of such policies. Finally, such an institution would be well placed to support work on the science of science and innovation policy along the same lines as pursued by the US National Science Foundation (NSF). In fact, the Observatory could function, as the NSF does, as a granting organization as well as a centre of policy and data analysis.

A final observation is that the NEPAD Office of Science and Technology is well set to support the experience of countries in the development of indicators and, in due course, this should lead to manuals or guidelines that deal with measurement in Africa. These Africa-specific guidlines would be in addition to those provided in the Frascati and Oslo Manuals. As the NEPAD Office of Science and Technology is now part of NESTI at the OECD, it could also contribute directly to annexes, or revisions to annexes, needed to interpret the manuals for use in African countries.

Asia-Pacific Region

In the Asia-Pacific Economic Cooperation (APEC), the Industrial Science and Technology Working Group (ISTWG) deals with innovation and indicators and there is work on innovation in life science and in small and medium-sized organizations (SMEs). In 2008, the ISTWG organized the APEC Symposium on Research and Innovation in Viet Nam which brought together government officials and policy-makers responsible for innovation strategy. The aim was to identify science and technology priorities and apply existing methods for assessing the priorities, and to build partnerships between government, the academic sector and industry. APEC has also produced the APEC Digital Prosperity Checklist which deals specifically with information and communications technologies but which contains some components of an innovation strategy.[3]

Of the 21 APEC members, Australia, Canada, Korea, Japan, New Zealand and the US are members of the OECD, and China and Russia are observers at the OECD Committee for Scientific and Technological Policy. The Association of South East Asian Nations (ASEAN) has the ASEAN Action Plan on Science and Technology, 2007–11[4] which focuses on science and technology capacity building for its ten member countries.

POLICY LEARNING

Country Reviews

The measurement of the activity of innovation is not an end in itself. Measurement is made to inform public policy debate and the development of evidence-based policies that bear on innovation. However, statistical measurement is not the only tool for policy analysis and review. The OECD conducts the OECD Reviews of Innovation Policy on the invitation of member and other countries. These reviews, involving a team of experts, support from the host government and an experienced

Secretariat, provide insight into the strengths and weaknesses of innovation systems and offer recommendations for their improvement. In Africa, an example is the review of South Africa (OECD 2007b), and in Latin America there is the review of Chile (OECD 2007e). The review of China (OECD 2008i) is an example of assessing an innovation system of a large emerging economy.

The World Bank, the United Nations Conference on Trade and Development (UNCTAD) and the United Nations Educational, Scientific and Cultural Organization (UNESCO) also conduct country reviews, as does the International Development Research Centre (IDRC), and there has been discussion of cooperation on such reviews of innovation strategies (Gault and Zhang 2009) to provide the same service to a wider range of countries as innovation and development policies become more interrelated. The reviews provide an opportunity for the host country to learn about the effectiveness of existing policies and to respond to the recommendations for change. As the OECD reports are published, any interest groups in the country involved are in a position to offer comments to their government.

Converting Knowledge to Value

In January 2009, the OECD and UNESCO, supported by IDRC and SIDA, convened a workshop to examine innovation in the context of development, with a focus on the combination of existing knowledge to create new knowledge and to convert that to value as part of the innovation process (UNESCO 2009). A broad view of knowledge was considered, going beyond that generated through formal research and development (R&D) processes, and the workshop addressed the question of innovation that happens without R&D.

This was a timely workshop on innovation, coming as it did in the midst of the financial crisis. The OECD Deputy Secretary General, Pier Carlo Padoan, in welcoming participants, reminded them that: 'it was innovation in financial services, and the rapid and global diffusion of the resulting products, that had caused the current financial crisis when these products lost value'; and he emphasized the importance of the concepts and definitions in the Oslo Manual (OECD/Eurostat 2005) that are needed to guide the discussion, and reminded participants that the Annex to the Oslo Manual, interpreting the manual for its use in the development context, was an example of a previous OECD–UNESCO collaboration (Gault and Zhang 2009).

Responding to the question of how to advance the role of innovation in development, participants noted the need to include reference to science, engineering, technologies and innovation in the Poverty Reduction Strategy Papers[5] (PRSPs) prepared by developing countries, to include innovation in the agenda of the OECD Development Assistance Committee (DAC) and to have a strong section innovation in development in the OECD Innovation Strategy when it was published in June 2010. The recommendation about the integration of science and technology into development policies was a reaffirmation of earlier work at the OECD which also addressed sustainable development (OECD 2006c, 2007f). Related measurement issues had been discussed by Bordt et al. (2007) and Gault (2007b).

From the measurement and analysis perspective, there was recurring reference by workshop participants to the need for case studies of innovation activities in developing countries, especially those that included innovation in the informal economy.[6] The work on innovation and R&D surveys in 19 African countries managed by the NEPAD Office of Science and Technology was seen as an example of capacity building, leading to an informed community of practice able to share knowledge on science, technology and innovation activities with statisticians and analysts in other African countries, directly and through the publication of the first *African Innovation Outlook* in 2010. Country reports were seen as a means of policy learning and improvement.

Innovating out of Poverty

Following the January 2009 workshop on 'Converting Knowledge to Value', there was a workshop in April 2009 on 'Innovating Out of Poverty' as a contribution to the OECD Horizontal Project on Food, Agriculture, and Development and to the OECD Innovation Strategy. In the Preface to the summary report (OECD 2009h), the Director of the Development Co-operation Directorate, Richard Carey, made the point that: 'There is a particular need to get innovation on to the development agenda and process, as well as to promote co-operation between developed and developing countries to achieve this'. This echoed the views of the participants in the earlier workshop.

The summary report of the workshop was prepared by the workshop Chair, Calestous Juma, and placed agriculture in developing countries in the centre of a global knowledge economy, requiring new management skills, institutions, policy coherence and international cooperation to achieve economic and social goals. This was a radical shift from viewing agriculture as a local and backward activity, and it requires extensive use of information and communication technologies (ICTs) to support the flow of data, information and knowledge needed to make it happen. The report went on to set an agenda for heads of state in developing countries which, if followed, could make innovation central to agriculture in developing countries.

The Bottom and the Middle

The United Nations Industrial Development Organization (UNIDO 2009) report makes a number of observations relevant to innovation and development projects. In summary, the observations are that industrialization leads to poverty reduction and to the achievement of the MDGs. Industrialization here is manufacturing, and manufacturing generates formal jobs which are better than self-employment or informal economy jobs as they are more able to support knowledge accumulation, are more secure and pay better than the alternatives. The report then suggests that climate change will adversely affect agriculture, but not manufacturing, which argues for a structural change from agriculture to manufacturing. In support of this change are opportunities resulting from globalization such as the extended value chains in manufacturing which allow firms to capture one part of the chain rather than having to acquire the ability to compete on the basis of the fully integrated production process. Globalization also means that manufacturing is dominated by trade and not by local markets, with implications for economies of scale and scope, but with a need for the appropriate infrastructure to support trade.

Industrialization is seen as 'lumpy', or inhomogeneous, in product range, space and time. The products, as suggested earlier, are evolutionary rather than radical in their change and, as noted in the Oslo Manual (OECD/Eurostat 2005), they involve both hard and soft technologies. The inhomogeneity in space recognizes the trend towards urbanization in the world and the advantages to be gained from co-location of production facilities in cities. The report points out that in regions of small countries, the creation of one city of sufficient size to support manufacturing clusters may cause tensions. The time inhomogeneity is between incumbents engaged in manufacturing and firms in countries that have yet to enter the global value chain. New entrants may not have access to the infrastructure of a large city, or the skills base, and there is a threshold that must be crossed before they are viable. This leads to the policy recommendations of the report.

There are two classes of countries considered: the low-income countries which must cross a significant threshold before their firms can enter world markets and be competitive; and the middle-income countries which are producing goods for the world markets but which face stiff competition. A government role for the low-income countries is stressed, as the market is not strong enough to prevail. The areas for policy action are infrastructure and framework condition improvement (Chapter 6) and support for naturally favoured locations to become agglomerations supporting the evolution of clusters. For middle-income countries, the suggested policy intervention is support for innovation, although the word is not used. The form of support proposed is knowledge generation through the creation or upgrading of technical and university education in cities. In addition, framework conditions are recommended that are conducive to easy entrance and exit of firms to ensure a Schumpeterian Mark I regime of creative destruction, leading to the flow of new ideas and the elimination of those not commercially viable.

Agriculture or Manufacturing?

Dealing with agriculture as a knowledge-based industry in a global economy and supporting manufacturing in low- and medium-income countries are not mutually exclusive, and could be symbiotic, especially if some manufacturing products use agricultural inputs. Both agriculture and manufacturing behave differently in a rapidly changing global economy. Both require an integrated policy approach involving several parts of government to provide the infrastructure and the appropriate framework conditions for these sectors to thrive. For the middle-income countries, the policy intervention in support for innovation will draw upon some of the components discussed in Chapter 7. For the lower-income countries, public policy must compensate for the lack of a mature market.

What about Services?

A characteristic of developed economies is that market services account for more than half of gross domestic product (GDP). When public services such as education, healthcare and government are added, the service component of GDP, and of employment, rises to over 70 per cent. The presence of mobile communications technologies in developing countries has supported knowledge transfer, allowing community producers to monitor prices at local markets, and the use of financial services, permitting the transfer of money using mobile phones. Micro-finance activities provide another example of an emerging financial services sector. There are also the services related to goods such as wholesale and retail trade and transportation and storage, which are necessary to move goods to markets.

With the importance of agriculture, extractive industries and manufacturing, there is considerable potential for policy initiatives to advance the relevant ancillary services. Of course, this needs a well-developed and well-connected workforce, with implications for education, training and lifelong learning policies. Not all jobs in service industries depend on a highly qualified, connected and mobile workforce. There are many jobs in retail services and personal services that make economies work, and the people in these jobs, and their willingness to provide service, contribute to making the country a desirable place in which to invest. Making services work is an area of potentially high return for innovation policy, especially non-technological innovation policy.

SUMMARY

This chapter has demonstrated that the tools for measuring the activity of innovation in support of evidence-based policy are available and well supported internationally, but the challenge lies in using them in developing countries, and documenting their use. The OECD provides a forum for the discussion of innovation measurement through its Working Party of National Experts on Science and Technology Indicators (NESTI) which includes as observers the NEPAD Office of Science and Technology from Africa, RICYT from Latin America and the Caribbean, Russia, China and Israel.

From a policy perspective, the components of innovation policies given in Chapter 7 can also be applied in developing countries, but there is a need for greater emphasis on building infrastructure to encourage economic development and to support the flow of knowledge. Development is path-dependent and this is more evident in a time of rapid global change; Beattie (2009) provides the example of the divergence of Argentina and the United States from comparable starting positions. Urbanization in Africa is providing opportunities for new industries and markets (Kapstein 2009), but effective use of urbanization in support of innovation and cluster development has implications for regional policies.

To support both measurement and evidence-based policy, more empirical work is needed of the kind initiated by the NEPAD Office of Science and Technology and supported by SIDA, case studies of the activity of innovation advocated by participants in the UNESCO–OECD workshop on Converting Knowledge to Value, and the building of communities of practice able to criticize and contribute to innovation policy development. This last point requires more government officials, and academics, with the capacity to work with innovation policy and indicators.

It has been impossible, in a single chapter, to do justice to the role of innovation strategies in development. There is a place for a book devoted to innovation strategies in developing countries with, perhaps, a single chapter making the link to work in the developed countries and noting the differences and similarities. The reader is also referred to the new *Handbook of Innovation Systems and Developing Countries* (Lundvall et al., 2009).

NOTES

1. www.un.org/millenniumgoals.
2. AMCOST was initially a 'Council'. In 2007 it became a 'Conference'.
3. The APEC Digital Prosperity Checklist is referred to in the APEC 2009 report on *Achievements and Benefits*, http://www.apec.org/apec/about_apec/achievements_and_benefits.html.
4. http://www.aseansec.org/8504.pdf.
5. The Poverty Reduction Strategy Papers are described at http://www.imf.org/external/NP/prsp/prsp.asp.
6. The United Nations University Maastricht Economic and Social Research and Training Centre on Innovation and Technology (UNU-MERIT) has issued a Call for Proposals for case studies of innovation and its measurement in Mozambique, Rwanda and South Africa. A complementary call to support graduate students in the doing of case studies as part of their research is to come from the Tshwane University of Technology, the Institute for Economic Research on Innovation (IERI). Both projects are supported by IDRC in Canada.

10. New directions

INTRODUCTION

This chapter presents some medium-term and short-term activities to advance the understanding of innovation strategy development, implementation, evaluation and learning. Chapter 11 provides some tasks for those engaged in the activities.

MEDIUM TERM

This section identifies six themes for more policy or measurement development. They are: analysis of the existing microdata on innovation; innovation without research and development (R&D) and user innovation; learning and failure; network analysis and complex systems; public sector innovation; and the science of innovation policy. They are chosen for special attention because of their potential impact on the understanding of innovation, leading to the possibility of finding more effective ways to promote innovation and related economic and social objectives.

In September 2006, 250 people from 25 countries gathered in Ottawa for the second OECD Blue Sky Forum, the previous one having been in Paris in 1996 (OECD 2001b). There had been previous 'blue sky' meetings on indicators at the Organisation for Economic Co-operation and Development (OECD), reviewed by Colecchia (2007), but the 2006 forum provided a place to bring together discussions of a systems approach to understanding a global, complex, dynamic and non-linear innovation system. This led to proposals for new work with the hope of insights into how parts of the system worked.

Participants put the case for moving innovation analysis from the study of activities, such as R&D and innovation itself, towards linkages, outcomes and impacts, while not stopping the decades of work on measuring activities (Gault 2007a). They called for greater cooperation across countries and with international organizations to improve the comparability of analysis, and for access to microdata in order to study linkages and outcomes of activities at the firm level. Arundel (2007) put the case for a better understanding of innovation in firms that perform no R&D, a topic he has raised elsewhere (Arundel et al. 2008b; Arundel et al. 2008a), and von Hippel (2007) showed the important role users play in the innovation process, a subject which over the years has received little attention in official statistics (Gault and von Hippel 2009), in analysis and in policy development. Finally, Marburger (2007) argued that the equivalent of the Minister of Industry, or of Science and Technology, should receive advice comparable to that received by the equivalent of the Minister of Finance, based on complex and intimidating models. For this to happen, he proposed a new cross-cutting social science, the science of science and innovation policy. This was a seminal meeting.

Since the Blue Sky II Forum in 2006, there has been rapidly growing interest in public sector 'innovation'. Some have been tempted to dismiss it as a misuse of the word 'innovation', which requires a connection to the market, but the other view is that the existing machinery of measurement, indicator development and policy analysis can be readily applied to the public sector now, and should be in order to understand how better to produce and deliver public goods and services, and how to benchmark and evaluate those processes. This work raises sufficient conceptual implications for the definition of a market and for the development of future manuals and guidelines, including future editions of the Oslo Manual, for the topic to be included here as an area for future work.

Microdata Analysis

The recommendation for microdata analysis was one of the outcomes of the OECD Blue Sky II Forum and it led to an OECD project, with support from a number of member countries, which has now produced a substantial set of findings. This work has been productive and should be continued with more elaborate data sets.

The microdata analysis project solves a number of problems. For the purpose of international comparison of the results, the project avoids having to deposit microdata with any agency outside of the country where the data were created. In many countries, where surveys are managed by a statistical office, legislation forbids the removal of microdata from the office as part of protecting the confidentiality of the respondents.

In the past, attempts have been made to produce anonymous data files which can be released, but this has not proved as successful with business data as it has been with social data. The reason for this is that the characteristics of people – age, gender, education, income, location – are, with the possible exception of income, well bounded and the likelihood of disclosure once identifiers are removed is small. This is not the case with firms, the characteristics of which are not randomly distributed about a mean value. An example is number of firms in size classes. Most firms (over 90 per cent) have fewer than ten employees in industrialized countries and the numbers fall off rapidly as size increases. Large firms, with 250 or more employees, account for less than 0.5 per cent of the total number of firms but will account for close to half the value added. These numbers, of course, vary from country to country but all countries exhibit the same characteristics, which are not unlike the relationship between number of cities in a country and their size, which behaves like the power law introduced in Chapter 1. When dealing with the large firms it is difficult, if not impossible, to produce anonymous data.

If it is agreed that the microdata analysis can take place inside the national statistical offices, or research institutes,[1] that hold the data, the next step is to agree on the econometric methods to be used to do the analysis, and then to ensure that the data have all been edited and imputed[2] in the same way. This last point is critical for international comparisons and may be one of the most important outcomes of this project: the encouragement of standard practice, not just in the use of concepts and definitions in respect of innovation, but also in the cleaning of the data and the preparation of final data sets for analysis.

Within statistical offices, in some cases, it is possible to enhance the data gathered by single surveys by linking the data gathered to data from other surveys and to administrative data. If this is possible, it frees the innovation survey from having to ask about employment, revenues or R&D expenditures, as this information can be added later and the burden of the survey is reduced. The use of data linkage, and of administrative sources such as tax or immigration data, are sensitive topics as there is concern about the collection, in one place, of large amounts of information about firms (or people).

In its first publication of the results of the OECD Innovation Microdata Project (OECD 2009b), there is a wealth of new insights which are only possible as a result of microdata analysis. The project examined: the determinants of innovation and the impact of innovation on productivity; modes of innovation, including non-technological innovation; and the incentive effect of intellectual property rights on innovation. The data used came from the fourth round of the Community Innovation Survey (CIS), or from national surveys outside of Europe that were close to CIS 4.

The project used 20 indicators to compare five dimensions of innovation: technological innovation; non-technological innovation; innovation inputs; innovation outputs; and a set of policy-relevant characteristics (internationalization, collaboration and intellectual property rights). The results of the analysis demonstrated considerable inhomogeneity of firms.

Size, as already discussed, was a major factor, with countries showing differences in respect of the introduction of product or process innovation for small and medium-sized enterprises (SMEs), but with greater comparability for large firms. Organizational and marketing innovation also varied significantly across countries.

Composite indicators were introduced to get at some of the comparability problems raised by Arundel (2007). An example was the combination of the novelty measure (new to the firm or new to the market), the international measure (Is the market of the firm international or domestic?), and an internally generated measure (Is the innovation based on in-house effort or not?). This allowed the identification of innovative firms producing innovations new to the market, sold internationally and based on in-house effort.

At the end of the analysis, there was still a need to understand why some firms innovate and others do not, a question that has preoccupied some Canadian policy analysts (CCA 2009a, 2009b) who see innovation as one component of a business strategy. There was also a need to learn more about the effect of innovation on the performance of firms. This is an issue of dynamics which calls for panel data. As indicated in the OECD report, panel data would require major changes in the sampling procedures in participating countries. It would also require a greater commitment of resources, which would only happen if innovation is seen as a key policy issue. Earlier in this section, reference was made to the lack of use of innovation indicators in the policy process, and one of the reasons for continuing and refining microdata analysis of innovation is to make innovation indicators, including those based on panel data, more central to the policy debate.

Innovation without R&D and User Innovation

One of the more robust findings from innovation surveys is that, within the survey universe, there is a higher propensity to innovate than to do R&D. Not surprisingly, this has been found in the OECD Innovation Microdata Project (Block and López-Bassols 2009: Tables S3 and S13). The R&D propensity in the Microdata Project is derived from innovation surveys and in Chapter 4 there was a discussion of why this figure could be larger than the R&D propensity measured in R&D surveys. The point here is that the gap is significant and it is almost certainly larger than that inferred from the Microdata Project.

This means that there are a large numbers of firms that innovate without doing R&D, and this raises various policy issues. Should these firms be encouraged to do R&D to narrow the gap, or should firms that do not innovate at all be encouraged to innovate, without engaging in R&D, with a resulting increase in the gap? Both could be considered, but more knowledge is needed about firms that innovate without doing R&D, a point made by Arundel (2007).

As discussed in Chapter 4, this gap exists in Canadian innovation statistics, and has done since the 1993 statistics Canada innovation survey and, since the programme was revised after 1985, there has been a more or less unchanged tax incentive programme to encourage R&D in Canada, the Scientific Research and Experimental Development (SR&ED) programme. However the SR&ED programme only accepts R&D in the natural sciences and engineering, including software R&D, but not in the social sciences. This means that firms that are doing R&D in business practices, organizational change or market development are not eligible for the tax benefit, and that might account for part of the gap.

Another part of the gap could result from firms engaged in technology adoption and user innovation. Recall that a firm that introduces a new or improved process that is new to the firm is an innovative firm.[3] One way of introducing a new or improved process or production technology is to purchase it and use it. No R&D is involved. A firm that adopts and then adapts a production technology is certainly innovative, and this is a case of user innovation which may take place without the doing of R&D. There is a third case where firms develop production technologies to solve their problems. This happens when they cannot find the solution they need offered on the market. These are also user-innovators, but they tend to be larger firms and they are more likely to engage in R&D. This has been observed by Gault and von Hippel (2009). User innovation, without R&D, is an example of what Lundvall (2007) would call learning by doing, using and interacting (DUI mode) as opposed to a science-based research process (STI mode).

The direct policy intervention may be tax incentives that encourage any R&D that would meet the definition in the Frascati Manual. An indirect approach, which might broaden the gap, is to encourage innovation and risk-taking as part of the education of the labour force with a view to building a culture of innovation. SME support programmes, like the Canadian National Research Council Industrial Research Assistance Program (NRC-IRAP), are able to encourage firms to grow[4] as a result of innovation. As larger firms have a higher propensity to do R&D, this could result in more performance of R&D in the business sector.

To gain a better understanding of innovation, and the performance of R&D, there is a place for panel data that could follow a population of firms for a number of years to observe how a population of non-innovative, non-R&D performing firms; innovative but not R&D-performing firms; and innovative and R&D-performing firms behave over time.

Learning and Failing

Innovation is about interacting within the firm, and with other firms and institutions, in order to create value. In the case of firms the value is determined by the market, but value can also be created in public institutions and that is discussed in the next section. The interaction, leading to the creation of value, is part of a learning process and it gives rise to knowledge accumulation and use. Lundvall (2007) would distinguish between two modes of learning, the DUI mode (learning by doing, using and interacting) and the STI mode (learning through science-based research processes). However acquired, it is the knowledge, converted to value, that permits the firm to survive and to learn from how it survives.

Dierkes (2001) has grouped firms into three broad categories. At the lowest level are those that take inputs, transform them into outputs, are driven by a price signal and operate in present time. So long as they can go on doing this, they can survive, but if inputs (people, energy and materials) become a problem because of economic shocks, or the market no longer values the good or service they produce, perhaps because of changes in regulation or behaviour, they are gone. A second level of firm operates like the first, but it has the capacity to monitor the production process, the acquisition of inputs and the marketing of outputs, and make changes which could be organizational, market related or a technologically different way of transforming inputs to outputs. It still operates in present time, but it can differentiate its product line and seek more cost-effective inputs. Then, there is the third level, which has all of the capacities of the second, but which adds a corporate memory and a foresight function and has the intellectual, financial and physical resources to transform the firm so that it can avoid the problems of the past and take advantage of the opportunities offered by the future. Size is a factor here, as is the kind of learning that goes on. The third-level firm can manage a mix of DUI and STI modes of learning and has the analytical capacity to take decisions based on past experience, which means that there is a memory of the past and a capacity for organizational learning (Dierkes et al. 2001b). The second level of firm is driven by the DUI mode and it is here that user innovation through modification of technologies and practices could be the dominant means of survival.

Organizational learning (Lundvall 1992) occurs in groups, in organizations such as firms or government departments, in regions and in countries (OECD 2002a). The learning activities and the knowledge accumulation support the activity of innovation in the countries, the regions and the organizations, and there is still much to be learned about the relationship between innovation and organizational learning. While this is important for future work in developed countries, it is even more important in developing countries where the capacity to learn as a group, as a firm or an institution has to be built before the capacity can be used.

While organizational learning can help solve problems and support innovation, it cannot solve all of the problems and there are points when firms fail. It may be that this is a significant learning experience for the people in the firm that can help them to move on and form new firms that will succeed. Such knowledge could be accumulated in venture capital firms that have worked with an industry for some time and which have learned from the failures of firms that they have supported. From the perspective of official statistics, there is little or no information available on failed firms. Assuming there is a business register and it is responsive to changes in the population it describes, the firm will have been removed and can no longer be surveyed. In Canada, there is a project run by the Impact Group to interview former managers from failed firms and to draw inferences for use in public policy development. This is a work in progress, but it does raise the issue of studying failure as well as success in innovation (Barber and Crelinsten 2009). This is an important consideration for future work.

Network Analysis and Complex Systems

A recurring theme in the book has been the global, complex, dynamic and non-linear nature of the innovation system, and reference was made in Chapter 2 to work on modelling complex and dynamic systems. This has a place in the science of innovation policy discussed later in this section on the medium term, but is also a subject in its own right which could shed light on the function of innovation systems. This is more a long-term, rather than medium-term activity, but it is an area for future work.

The OECD explored this by running a workshop on innovation and networks which may lead to further work resulting from the Innovation Strategy. The American Association for the Advancement of Science (AAAS) has produced a special report on complex systems and networks (AAAS 2009), including an paper on predicting the behaviour of techno-social systems (Vespignani 2009) which does have implications for innovation analysis.

Public Sector Innovation[5]

While innovation in the public sector has been discussed for years, and there are studies and journals[6] devoted to the subject, it has more recently become of greater interest in international organizations. This section looks at innovation in the public sector and where that work could go.

There are two aspects of the public sector that enter a discussion of innovation. The public sector is the provider of framework conditions and infrastructure which support the activity of innovation in the private sector. The public sector also manages activities that provide goods and services to people or to other parts of the public sector. Examples are social services and benefits delivered to the unemployed, and analysis and policy advice commissioned by, or generated within, departments of government and used to change or initiate legislation and regulation.

Framework conditions and infrastructure were discussed in Chapter 7 as components of innovation strategies, and again in Chapter 8 as matters for coordination as part of supporting innovation. This leaves innovation as applied to the activities of the public sector for consideration here. The first question to address is whether the term 'innovation' can be applied to the public sector.

In Chapter 3, the current definition of the activity of innovation provided by the Oslo Manual (OECD/Eurostat 2005) was given, and each of the four components of the definition – product, process, organization and market innovation – connect directly or indirectly to the market. Producers of public goods and services do not connect to the market. However, that is not the end of the discussion.

Chapter 3 also listed a number of innovation activities, such as R&D, capital investment, training and knowledge acquisition, which may form part of the activity of innovation. Not all are needed for innovation, a point made earlier in this chapter when discussing innovation which happens without R&D being done in the firm. The same innovation activities can, and do, take place in public institutions, governments, hospitals, other healthcare service providers, and in institutions of education. This suggests that a first step towards measuring public sector innovation would be to take the Oslo Manual and to measure the resources allocated to innovation activities, and then to use the resulting indicators to support discussion of priorities for resource allocation (What should the public institution be spending its resources on?). This raises the question of why the activities of innovation are undertaken and how the outcome is evaluated.

A firm seeks energy, materials and people in the most cost-effective way to support the transformation process leading to product delivery to the market. The objective is to get the product to market and to make money. In the public sector there is a procurement process, but it may also be part of broader policy objectives, such as supporting SMEs, developing regions or being a demanding client in order to improve products being offered in the country. Similarly, the transformation process leading to the provision of education, healthcare and social services may involve policy objectives that go beyond getting products to the client in the most effective and efficient way. Examples of such policies are employment equity, language policies in multilingual countries, or regional development imperatives. Finally, the products are not sold at a profit, and they may be supplied by monopolies, which suggests that the analysis of public sector innovation has to be quite different from that in the private sector.

To add to the problem, the differences between the private and public sector are not always clearly defined. The use of public–private partnerships to achieve public objectives in a businesslike way mixes the two. There are cases of privatizing what have previously been public sector activities, such as prisons, waste collection and security services, while retaining control through the contract, its finite term and the competition at the end of the term. There are also public takeovers of firms, as seen in the cases of financial institutions in some countries in 2009.

Assuming that the Oslo Manual could be adopted for use in the public sector, the next step would be the development of a questionnaire comparable to the Community Innovation Survey questionnaire for use in the private sector, followed by the production of indicators as an input to the management of public resources as part of the innovation process. One of the desired outcomes of public sector innovation, especially in times of recession, is the saving of public funds so that taxes can be reduced or the resources can be reallocated to higher priorities of the public institution. While such an outcome could be measured, in the spirit of the Oslo Manual, attention should be paid to the longer-term impacts of the savings. Following the recession of the early 1990s, the Canadian government established priorities, reduced the public service and moved the federal budget from deficit to surplus, a state that lasted until the 2008–09 financial crisis. In the course of doing this, departments lost their analytical capacity while protecting their various lines of business. The reduction in the policy analysis and development function had to be remedied some years later with the establishment of the federal Policy Research Initiative.[7]

The next step for measuring public sector innovation and using the results to inform public sector reform is the development of guidelines and definitions that can be used to gather information from public institutions. In the longer term, attention will have to be given to the public sector functions not discussed in this section, but which are dealt with in the Oslo Manual and which are part of private sector innovation: the framework conditions and the provision of infrastructure. The question then is whether to merge public and private sector manuals in order to have one innovation manual, or not.

Manuals provide the language to talk about phenomena, such as innovation, but language is contextual and influenced by culture. The culture of innovation in the public sector differs from that in the private sector. The counter observation is that there are parts of the private sector that are averse to innovation, and parts of the public sector that are able to take risks and to make changes.

There are enough issues to support a discussion of how to extend the language of discourse on innovation, in a meaningful way, to include the public sector.

The Science of Innovation Policy

Part III reviewed some components of innovation policies (Chapter 7) and then considered how selections of the components were coordinated by some governments to produce innovation strategies. As components can include framework conditions as well as more direct interventions, and coordination can take place at many levels of government, the problem of developing an innovation strategy is complex, as is predicting or even understanding its impacts. To improve the understanding of the consequences of policies in the area of innovation, there have been calls for the creation of a new social science to study the problems.

The focus here is on innovation and on innovation policy; however there is also interest in the science of science policy as well as the science of science and innovation policy. It was John Marburger, the Director of the Office of Science and Technology Policy (OSTP) in the administration of US President George W. Bush, who in 2005 called for a science of science policy (Marburger 2005) and went on to put the case for a science of science and innovation policy at the OECD Blue Sky II Form in Ottawa in 2006 (Marburger 2007). Since then, the US National Science Foundation (NSF) has managed three solicitations on the subject of the Science of Science and Innovation Policy (SciSIP) and in March 2009 there was an NSF–AAAS workshop[8] of SciSIP grant holders to assess progress and to build a community of practice around SciSIP as an emerging discipline.

Meanwhile, *The Science of Science Policy: A Federal Research Roadmap* (NSTC 2008), a report to the Sub-Committee on Social, Behavioral and Economic Sciences of the National Science and Technology Council (NSTC), was released. It set out a science of science policy roadmap that included three themes: understanding science and innovation; investing in science and innovation; and using the science of science policy to address national priorities. This is an important document in the context of the US federal government and for the direction of innovation policy. A second document laid out the direction of *Social, Behavioral and Economic*

Research in the Federal Context (NSTC 2009). In NSTC (2009), innovation and creativity are linked together throughout and, after a discussion of the 'complex ecosystem of innovation', the observation is made that:

> Efforts in these areas of the human sciences are essential for understanding how innovation systems work. These efforts will lead to better monitoring of educational outcomes, financial returns to R&D and the innovation life-cycle, as well as better ways of monitoring and evaluating the outcomes of our nation's public and private R&D efforts. One component of this effort is the development of an interagency 'science of science policy' task group that is preparing a report on this emerging science, with a focus on innovation. (NSTC 2009)

Both documents are clear on the importance of understanding how the innovation system functions. However, they focus on a 'science of science policy' rather than on the 'science of science and innovation policy' as advocated in Marburger (2007) and as supported by the NSF in its SciSIP programme. There is also little on the science of policy as the documents are more concentrated on science policy as managed by the US federal government.

There is still a need to respond to Marburger's call for cross-disciplinary work on understanding how innovation policy works or does not work. Of course, this presupposes that there are well-understood objectives for innovation policy against which to judge its effectiveness. Some of these objectives were discussed in Chapter 6 and again in Chapter 8. Given that not all of innovation is tied to science, there is a case for examining innovation policy, as an academic subject, separately from science policy or R&D policy. From the perspective of science policy in the US, questions have been raised about its existence (Sarewitz 2003).

A science of innovation policy could include taxonomy of the possible components of an innovation policy, separating them into areas for direct policy intervention and the framework conditions that help the system to work, such as education, healthcare, good governance, financial services, and transportation and telecommunications infrastructure. A second domain would be the understanding of the coordination of some of the components by various levels of government and across different institutions. Finally, there is the question of how well the system works, which raises political science questions about institutions, standards and interactions; sociological questions about communities of practice and the learning capacity of groups, institutions and regions; and economic questions about growth, employment and priorities for resource allocation. The engaged reader intent on including all of the social, behavioural and economic sciences could find questions that can be addressed by the disciplines of geography, cultural anthropology or criminology.

While there has been work since the 1980s on the understanding of innovation systems, the next challenge is to understand how innovation policy works. One of the shifts in emphasis will be from the natural sciences and engineering to the social, behavioural and economic sciences, as they have the machinery to deal with the human and institutional learning and interactions which are part of the innovation process. Another shift, in a science of innovation policy, is from basic science and the commercialization of new knowledge to focusing on turning existing knowledge into value in an effective way.

While the subject matter emphasis may shift, there is still a key role for statistical indicators of the activity of innovation, the linkages of the actors, the outcomes and the social and economic impacts. These were discussed in Chapters 3 and 5 and they are needed to monitor and to evaluate policies that affect innovation, including the framework conditions as well as the more direct interventions and the means of coordinating policy activities.

The science of innovation policy is a challenge for all countries, as all businesses in all countries are engaged in some form of innovation and are supported or constrained by national and international framework conditions. The NSF is supporting work on the subject, the European Union (EU) Seventh Framework Programme could be used to advance such work, and a recent study commissioned by the European Commission, Directorate-General (DG) Research reports on policy mixes for R&D in Europe (Nauwelaers 2009). While the policy mixes report does not deal explicitly with innovation, it does get close to the questions that must be answered, or at least addressed, by a science of innovation policy. In Africa, the African Union is considering establishing an African Observatory for Science, Technology and Innovation which could support work on the science of innovation policy in Africa, as well as being a repository for indicators and innovation policies, and a source of information and critical analysis.

SHORT TERM

This section addresses activities that can advance the development of indicators and strategies over the next two or three years. The principal directions coming out of the OECD Blue Sky Forum II are still present: the importance of being able to use new and existing indicators to tell a compelling story to the policy community; the shifting of emphasis to measures of outcomes and impacts, while retaining measures of activities and linkages; and the need for more microdata analysis to inform the policy process and the access to the data which makes the analysis possible.

The call for work on the science of science and innovation policy, issued at the Blue Sky Forum II, has been met by the US National Science Foundation and the present need is the building up of a community or communities of practice around the subject and more work on the science of innovation policy. If more effective support for innovation, leading to sustainable productivity growth, is the goal for both developed and developing countries there is a case for sharing the knowledge gained by practitioners.

Measures

If innovation in the public sector is to be treated as an object of innovation policy, indicators of the activity and its linkages to other parts of the innovation system have to be developed, measured and compared over time and across countries. This follows the same path as all other indicators developed by the OECD and codified in the Frascati Family of manuals.

The role of the user in innovation, and the different way users deal with intellectual property compared with producers, should be probed in more countries so that international comparisons can be made. The point in the text that there is already a strong signal of user innovation in the results of CIS surveys would suggest that these indicators could be developed by conducting some modest follow-up surveys to the last innovation survey conducted in the country.

Similarly, there is a case for more information on firms that innovate but do not do R&D. The motivation for this is to provide the support for policy to promote the innovation by these firms so that they are able to grow. As the propensity to do R&D is dependent on the size of the firm, the likelihood of these firms performing R&D would be expected to increase.

Analysis

Work on innovation strategies at the Commission and at the OECD was originally motivated by the opportunities and challenges offered by the emerging economies and their impact on world markets. Since then, innovations in financial services have diffused rapidly and widely, and their value has collapsed, with a major impact on economies and societies. The financial crisis has resulted in a response of governments and of governments working together through the G20. This public sector response to private sector initiative makes the understanding of innovation, and the frameworks within which it happens, even more relevant and urgent.

There are initiatives to provide working definitions of public sector innovation and to develop indicators to monitor innovation in the public sector. This is a new domain of the subject, but a natural extension of what has been going on since the 1980s. In addition, the economic crisis provides an opportunity to analyse and understand how the public sector has responded through market intervention, improved regulation and dynamic leadership. This underlines the point made in Chapters 2 and 8 about the public sector being an integral part of the innovation process.

The financial crisis of 2008–09 also provides abundant information on the social and economic outcomes and impacts of innovation in financial services and the rapid diffusion of the products offered. Understanding the process and the results is part of being able to tell the story, albeit after the fact, to the policy-makers, including the regulators, to avoid a repetition of the recent crisis. The knowledge would also contribute to the development of a science of innovation policy.

More generally, the importance of working with microdata to provide better and internationally comparable results has already been demonstrated as part of an OECD project (OECD 2009b). The project uses the same econometric model in each participating country and then expects each country to deal with its data in the same way in order to ensure comparability of the results of the analysis. This is a major step towards international comparability, as the methods of dealing with non-response to survey questionnaires, and of partial response, has not been the same for all collectors of official statistics. This is a standard that could and should go beyond innovation surveys. The second requirement for microdata analysis is access to the data. Not all statistical offices permit researchers from outside to work on confidential data. This is a matter for discussion in international forums.

Development

Chapter 9 covered survey and policy activity in developing countries and the link with the discourse on concepts, definitions, surveys and policies going on in international organizations and in the EU. The EU, the OECD, the United Nations Educational, Scientific and Cultural Organisation (UNESCO), the World Bank, and other organizations have all contributed to part of the discussion and it may be time to consider greater coordination of support for surveys, analysis and policy advice at a time when support to developing countries is at risk of being reduced in response to the financial crisis. However, the call for this should come from the developing regions as part of solving their own problems of promoting innovation and sustainable productivity growth.

Science of Innovation Policy

As discussed in the last chapter, there have been three rounds of solicitations and the beginning of a community of practice around the NSF Science of Science and Innovation Policy (SciSIP) activity. However, if SciSIP is to support innovation strategies and their implementation a needed next step is the appearance of the cross-disciplinary social science that will, through its work, deepen the knowledge of how innovation policy works and can be made to work better. The following section deals with the establishment of what could be a laboratory for the science of innovation policy.

Learning and Dialogue

More analysis, and the preliminary insights from a science of innovation policy, would help decision-makers to act to make innovation policy more effective, to advance toward sustainable productivity growth and to address global challenges. However, this work would be better focused if there were a secretariat to collect the outcomes of the analysis, to synthesize key findings, to propose more research, and to bring together regularly a group of decision-makers from the public and private sectors as a collective learning activity.

An immediate outcome of such a forum would be a broader understanding that innovation takes place in a global, complex, dynamic environment with a non-linear response to policy interventions, and that it involves human resource policy (education, training, lifelong learning) as much as science and technology policy and policies related to finance and trade. There is no one policy for all countries. However, there are common elements, and understanding how the elements work would support learning, and better implementation of policies. Ideally the policies would be implemented in a coherent manner, whether across the whole of government or in particular areas such as agriculture, energy or defence.

To be effective, the forum would have to function at an appropriate level. Looking to high-level forums for a model, there is the Davos World Economic Forum which is, perhaps, at too high a level. The people at the table discussing what works in innovation policy should be those that recommend policy to government ministers and those that are affected by the implementation of such policy in other public institutions, like universities and research institutes, and in the private sector. The group of stakeholders would be an international equivalent of the many high-level bodies of public and private sector decision-makers established to advise governments on innovation strategies and their implementation. The level suggested here is one below that of the head of a government department.

Once the level is established, there is a problem of the complexity of innovation and innovation strategy. This suggests that the same people would not be at the table for every meeting. A meeting on human resources to support innovation might have participants from education ministries as well as people developing immigration policy. If the topic were the promotion of innovation in firms that did no R&D, the participants would be responsible for SME policy and, perhaps, intellectual property policy. To make this happen, there would have to be a list of topics agreed at an early meeting and a secretariat able to bring the right people to the table.

That raises the question of the secretariat. For it to work, it would have to be led by a senior person of at least the same standing as the participants, and the staff would have to be expert in the subjects that are needed to support an innovation strategy.

Such a forum was called for in 2007 when the OECD Council, meeting at ministerial level, initiated the OECD Innovation Strategy (Chapter 6). The OECD would be the logical place for it and its secretariat, as staff and participants could also interact with the committees and working parties at the OECD which span all of the subjects relevant to innovation strategy.

SUMMARY

This chapter has proposed some medium- and short-term activities for the innovation policy, analysis and measurement communities. The proposals are there to stimulate debate in the various organizations that are trying to understand innovation strategy as a means of achieving economic and social goals.

Medium Term

This section has provided an agenda for work in the medium term that includes technical developments in support of microdata analysis, a better understanding of innovation in firms that do no R&D and the role of user innovation in those firms, and the importance of learning and the conversion of the resulting knowledge to value, whether as a private or public product. Expanding the subject of innovation to the public sector is encouraged, and the final challenge and a significant one is the development of a new and cross-cutting social science to improve the understanding of the science of innovation policy.

One of the recurring themes in the book has been the need to address dynamics in the analysis of innovation systems, and this means modelling the system in a way that supports dynamic analysis, and that means the acquisition and use of longitudinal data on the actors in the system and their interaction. Learning, knowledge accumulation and use are part of a dynamic process and it is a fundamental area for more work if the analytical community is to provide credible advice to policy-makers and contribute to the science of innovation policy.

Short Term

In the short term this chapter has made a number of proposals. They are the following:

- Measures:
  - Public sector innovation, concepts, definitions and statistics on innovation activities.
  - User innovation, statistics on sharing of intellectual property.
  - Innovation without R&D, statistics on firm behaviour.
- Microdata analysis of firms and of doctorate holders:
  - More countries.
  - Expansion of datasets by linking to administrative and survey data.
  - Discussion of greater access to confidential data, while following the rules of confidentiality used by national statistical offices to protect respondent information.
- Development:
  - Support for survey work, case studies and analysis.
  - Coordination of the work of international organizations and the EU at the initiative of developing regions.
- Science of innovation policy:
  - Work on the science of policy as well as on innovation policy.
- Learning and dialogue:
  - Establishment of a high-level forum on innovation strategy.

The time of recovery from the economic crisis is an opportunity to address the problems of promoting innovation in a global, complex and dynamic system that is non-linear in its response to policy interventions, as part of achieving the goal of sustainable productivity growth. The goal of sustainability will ensure that the global challenges of climate change; supply and security of energy, food and water; and inequality are part of the solution.

NOTES

1. In some statistical offices a new survey carries with it such administrative overhead as to make it practically or politically impossible to launch a new survey. This overhead can range from requiring the approval of the minister to having to have a law passed by the Parliament to give authority for the survey. When the overhead becomes high, other ways are found of collecting the data and one way is for the client for the information to contract out the survey and analysis to a research institute. This has advantages, in terms of timeliness, but research institutes do not have the legal authority to compel response. That is the prerogative of the statistical office. This means that the voluntary survey will have a low response rate which will call into question estimates of variables for the survey universe based on the responses received.

2. Survey questionnaires are sent to a sample of survey units. Not all come back. This is an example of non-response. Some come back but with not all of the questions answered. This is an example of item non-response. Even when there are responses, they may not be right. A simple example is reporting revenue of $100000 in a box which asks for number in the thousands. The erroneous response is $100 million, not $100000. Survey statisticians have ways of editing survey responses, and the programs should catch the $100 million if there is comparison with previous responses, or if there are ratios such as revenue to the labour force at the survey unit. On occasion the basic edits cannot resolve the problem and the respondent has to be contacted. The point here is that not all edit procedures are the same. It could be argued that this does not matter, so long as they produce a file with correct responses. Imputation is another matter. If there is non-response or item non-response, there are ways of imputing the responses. One way is to look at a comparable firm and use its response. Another is to look at a response from a previous period and to use that. Imputation is a useful statistical technique, but it should be well documented and the percentage of the population estimate accounted for by imputed data should be known as a data quality indicator. One of the outcomes of the Microdata Project is an examination of imputation procedures leading, ideally, to better international comparability.

3. Strictly, an innovative firm has engaged in the activity of innovation in the reference period of an innovation survey, usually the last three years. This raises some interesting questions. In some industries capital equipment has a long lifetime, longer than three years. This means that the firms that had purchased the leading-edge machines four years ago, or longer, will not appear in the innovation statistics.

4. Growth is not the objective of every firm. A study by Bordt et al. (2004) identified firms that, for various 'lifestyle' reasons, had no wish to grow. However they did welcome government support. This does raise the question of whether growth should be part of the policy objective for SME support programmes.

5. This section has benefited from the work of Ray Lambert and Carter Block and from considerable discussion with Stéphan Vincent-Lancrin. They are not responsible for the interpretation.

6. An example of a public sector innovation journal may be found at www.innovation.cc.

7. The history of the Policy Research Initiative may be found at www.policyresearch. gc.ca.

8. The details of the AAAS workshop are available at www.aaas.org/spp/SciSIP/.

11.   The roles of the players

INTRODUCTION

Chapter 10 provided an agenda for future work on activities to improve understanding of innovation strategies in the medium and short term. This chapter looks at the people who are going to do the work and makes recommendations for what needs to be done. The recommendations are based on years of experience and observation resulting from dealing with policy-makers and official statisticians in their own countries and in international organizations. The perspective, therefore, is from the working level where the objective is to get things done, specifically those things proposed in Chapter 10.

This is a book about innovation strategies in a global economy, their development, implementation, measurement and management. It assumes, explicitly, that innovation can be a power for good, but only if it is managed, and it recognizes that the system in which innovation occurs is global, complex, dynamic and non-linear in its response to the policy intervention. The global nature means that no government policy can control the entire value chain. Complexity means that innovation is not described by sound bites, the dynamics of the system may be impossible to manage, and counter-intuitive outcomes of government policy may result from the non-linearity of the system.

In the next few pages, the roles of the players in the policy system are considered. They are: the senior person from industry; policy analysts from government; academics and researchers; policy analysts from international organizations; representatives of civil society, including the consumer, along with industry and labour associations; and the official statisticians. The topic of development is covered by the players in the various categories. It is no coincidence that these actors could also be at the table as members of a high-level council advising the government on the development and implementation of its innovation strategy, and at international forums discussing the same issues from a global perspective.

INDUSTRY

The senior person from industry, the Vice-President Corporate Strategy for example, does not need government to tell the firm how to function. However, the firm does need working infrastructure (universal broadband access, a functional and secure transportation system), and framework conditions that are necessary to support business (regulation to protect society, a justice system, trade support), but are sufficient to avoid the excesses of the financial crisis. As a more basic example, meat packing plants, in most countries, do not put products on the market that kill people. There are regulations and government inspectors and inspections, leading to a public confidence in meat products.

As markets are global, the firm could benefit from the knowledge that a government can provide through trade or diplomatic missions. The question then is how to get the knowledge from the government to the creators of wealth as part of an innovation strategy. There is more to this than unidirectional knowledge transfer. It is an opportunity for firms and governments to learn from one another and to build networks that store the knowledge resulting from the learning opportunities. It is also an opportunity for firms to draw on this knowledge as part of developing their business strategies where innovation might be a component.

Firms and governments also interact through direct and indirect support programmes which address market failures such as tax support for research and development (R&D), and systems failures, of which the current economic crisis may be an example. Some programmes are more appropriate for large firms, such as local, regional and national government incentives to locate plants of multinational firms, others such as the US Small Business Innovation Research (SBIR) program, the UK Small Business Research Initiative (SBRI), or the Canadian National Research Council Industrial Research Assistance Program (NRC-IRAP) are directed at SMEs. The challenge is for the decision-makers in industry to see the bigger picture that not only leads to a more competitive environment, but which also embeds institutional learning in the innovation process, and moves the country closer to its economic and social goals.

The benefit to the private sector of getting involved in government decision-making is better framework conditions and more effective pro-grammes of government which affect the private sector and make it easier to benefit from innovation.

GOVERNMENT

The reasons for senior officials of government to get involved in innovation strategies were put in Chapter 1. There are global problems that have to be addressed, such as climate change, and there are urgent local problems in developed countries such as the ageing population, and there are problems looming related to paying for the government interventions to mitigate the financial crisis, as governments try to move closer to balanced budgets. Equally urgent is the need to support economic and social initiatives arising from developing countries. Governments need financial resources to deal with these problems.

These objectives cannot be addressed without political support, and the belief of government that innovation provides a path to achieving the objectives. This means that the government must be prepared to bring the need for an innovation strategy to the legislature and to the people. This requires briefing material, the management of interdepartmental meetings, and the preparation of draft legislation dealing with framework conditions or with more direct interventions that support innovation. This is a significant undertaking for the policy analysts involved.

Innovation provides potential for economic growth but it also requires the management of a scarce but renewable resource, the labour force, and all of the programmes that support it, including education and training, lifelong learning, immigration, healthcare and social safety nets. It also requires a cultural change to promote entrepreneurship and risk-taking, and to bring innovation and its successes, and failures, into the public discourse. Another shift needed is recognition of leaning by doing, using and interacting (Lundvall's DUI mode) leading to innovation in firms without doing R&D, to firms that engage in user innovation, and to consumers who innovate as a result of this knowledge (Lundvall 2007).

These activities, however urgent, and socially valued, are the responsibility of more than one government department. The challenge for the senior official is to work with other senior officials to produce an innovation strategy that addresses the problems and which can be implemented. This is a step away from saying that the implementation of the strategy should be managed as a whole-of-government initiative. It may be sufficient to conceive of the strategy as a whole-of-government activity and then to divide up its implementation in a way that preserves the authority of the departments that have to do the work. Then, there is the question how to manage the monitoring and evaluation activities, and the revision of the policy implementation or of the policy itself that should follow. These are operational problems that bureaucrats should be able to solve and they could also be research topics for academics in the new science of innovation policy raised in Chapter 10.

There is another challenge raised in Chapter 10 for the senior bureaucrat and that is innovation in the public service and, more generally, in public institutions. This is an opportunity for government to engage in innovation activities and to learn by doing. The knowledge gained can inform discussion with other players in the innovation system.

UNIVERSITIES AND RESEARCH INSTITUTIONS

There is an active community of scholars analysing systems of innovation at the national, regional and local levels and some of this work was reviewed in Chapter 2. There are papers in journals, conferences and institutes preoccupied with the subject. Some academics influence decisions in governments but, is this enough? Has the time come for a deeper understanding of the innovation process with findings that contribute to the strategies of business and government?

The challenge for the academic community is the participation in the linked initiatives of the Science of Science Policy (SoSP), of Science and Innovation Policy (SciSIP) or of Innovation Policy (SciIP), while continuing the work on understanding the activity of innovation.

INTERNATIONAL ORGANIZATIONS

Chapter 6 looked at the work of the Organisation for Economic Co-operation and Development (OECD) and of the European Commission on the development of innovation strategies for OECD countries. This is challenging work and some very good people are involved. The work has engaged member countries through OECD committees, and EU member states through workshops and consultations. The outcome of both processes will appear in mid-2010. It could be argued that whatever the outcome, the subject of innovation and the need for innovation strategies has been brought into the discourse of senior officials and this is a necessary part of building an innovation culture.

The development component of innovation strategies was discussed in Chapter 9 and there is a role for the development analysts to ensure that innovation becomes part of the development agenda and that the strategies include the involvement of people from developing countries and regions in the discussions.

A challenge for the OECD Innovation Strategy is to have at least the same influence as the Technology Economy Programme (TEP) of the OECD. It began in 1988 and produced a report (OECD 1992b) which, with the recommendations of a High-Level Group of Experts, provided the basis for a declaration concerning technology and the economy adopted by OECD Council, meeting at ministerial level in June 1991.

The TEP began after the financial crisis of 1987 and addressed the need for the understanding of the interactions between technological development, the economy and society, arguing that: 'An integrated and comprehensive approach of this sort is indispensible to informed policy-oriented decision-making' (OECD 1992b: 3). As Lundvall has noted, the TEP initiative of Robert Chabbal, Director of the Directorate of Science, Technology and Industry (DSTI), integrated the ideas of innovation scholars and gave innovation policy as well as innovation studies, a new kind of legitimacy (Lundvall 2007). In 2009, the stage is set for another boost to innovation policy, and a challenge to innovation scholars to go beyond understanding innovation to understand how innovation policies work once they are implemented.

While the focus has been on the work of the OECD and of the European Union (EU), there are United Nations (UN) organizations engaged in support for innovation, such as the UN Educational, Scientific and Cultural Organization (UNESCO), the UN Conference on Trade and Development (UNCTAD) and the UN Industrial Development Organization (UNIDO) and all of the organizations contributing to progress towards the Millennium Development Goals by 2015. There is also the World Bank and its programmes, especially those related to capacity building. There are also the philanthropic organizations that function not unlike international organizations. Given that resources are scarce, there is a need for the people in these organizations to talk to one another.

CIVIL SOCIETY, INDUSTRY AND LABOUR ASSOCIATIONS

Innovation affects people and their institutions. This suggests an active role for representatives of civil society, industry associations and labour associations in the innovation process, in policy development, implementation and evaluation. At the OECD, the Business and Industry Advisory Committee (BIAC) and the Trade Union Advisory Committee (TUAC) have observer status and take part in Committee debates. The question is the extent to which civil society, industry and labour associations are engaged in the innovation strategy process in countries.

OFFICIAL STATISTICIANS[1]

Most of the expectations of official statisticians were given in Chapters 4 and 5, they followed from the outcome of the OECD Blue Sky II Forum where the emphasis was on moving more to output and impact indicators to support telling a story to the policy community based on statistical measurement.

It is worth going back to the outcome of the OECD Technology and Economy Programme (OECD 1992b) to see the view then:

> The main general area for improvement lies in the work aimed at integrating statistics previously collected and analysed and published separately (especially science and technology statistics, industrial statistics and education and employment statistics). New indicators should be developed for innovation and its diffusion and for intangible investment and its components. More attention needs to be paid to data with an international dimension in order to contribute to a better analysis of globalization. There is a need to collect better data on human resources, especially data on training and the supply and demand of scientists and engineers. More controversial, given the rules governing the official collection of statistics, is the need for indicators on firms, especially MNEs [multinational enterprises]. The indicators of long-term research should be improved, particularly in the higher education sector. (OECD 1992b: 21)

The basic issues are the same, but much has happened in the intervening years to improve the understanding of innovation, of innovation systems and of innovation strategies.

SUMMARY

This book has covered a lot of ground, from statistical measurement and indicators, their development and their use, to the development and implementation and evaluation of innovation strategies, using the OECD and the work of the European Commission as examples.

The objective of putting all of these topics in one place is to show how they interconnect, or should, and are all part of an approach to building an innovation strategy. Having put the pieces together, the book ended with a work programme for the institutions engaged in building innovation strategies in developing and developed countries, and then went on to assign some homework to the reader who works in one of these institutions.

The final task for the reader is to provide the minister responsible for innovation with a one-page summary of the text in support of a process of developing, implementing, evaluating and learning, all related to a coherent innovation strategy.

NOTE

1. Official statisticians are responsible for official statistics. If the reader wishes to know more, there is the International Association of Official Statisticians website, http://isi.cbs. nl/iaos/.

Appendix A: Sources of information

INTRODUCTION

Much information in the text is drawn from various websites which have not been added to the References. In this appendix the reader is guided to the websites that have been used. The list is by no means exhaustive, but it covers most of the information sources used in the text and provides an opportunity for the reader to pursue related lines of enquiry.

INFORMATION SOURCES

Commission of the European Communities (CEC)

The CEC is the source for the papers relating to the EU Innovation Strategy discussed in the text. The starting point is: http://ec.europa.eu/. Select a language and then click on 'Who's Who', then on 'Directorates-General'. The starting point is DG Enterprise and Industry, go to 'Innovation', and then to 'Background documents'. That will produce most of the CEC references in the text.

In addition, there are the documents found on the Pro Inno Europe website, www.proinno-europe.eu. The Pro Inno Europe initiative combines analysis and benchmarking of national and regional innovation policy performance with support for joint initiatives by innovation agencies and other innovation stakeholders. The site provides the European Innovation Scorecard (EIS) and relevant analysis.

Community Innovation Survey Questionnaire

The generic questionnaire in English is taken from the OECD website, http://www.oecd.org/dataoecd/37/39/37489901.pdf.

Eurostat

Eurostat is the Statistical Office of the European Communities. Information on innovation and research can be found at epp.

eurostat.ec.europa.eu/portal/page/portal/structural_indicators/indicators/
innovation_and_research.

NEPAD OST

The New Partnership for Africa's Development (NEPAD) is a programme
of the African Union (AU). It has an Office of Science and Technology
(OST) which promotes the measurement of innovation. The history of
science, technology and innovation indicators in Africa was described
in Chapter 9, but there is more to the story and it can be found at www.
nepadst.org. This is the website for the African Ministerial Conference
on Science and Technology (AMCOST). Use the Document Library to
follow the history of innovation measurement.

NSF

The US National Science Foundation is well known for its biennial
indicator report which can be found at: www.nsf.gov/statistics/seind08/.
However, the website is a very rich source of information and should be
explored before clicking on 'Statistics' at the top and looking for indicators. What will not be found are indicators of the activity of innovation.
For these, the reader must wait.

However, while waiting, have a look at the available information on
the new Business R&D and Innovation Survey (BRDIS). This is found
on the website of the US Census Bureau as the Census Bureau does the
survey in partnership with the Science Resources Statistics (SRS) Division
of the NSF. Start with the URL www.census.gov/econ/overview/mu2600.
html, and for the questionnaire, click on 'Questionnaires' at the top of the
page.

OECD

The Organisation for Economic Co-operation and Development (OECD)
provides internationally comparable statistics, analysis and policy advice to
its 30 member countries, observer countries, and other countries and organizations involved in the global economy. The website is www.oecd.org.

On the website, 'Statistics' is on the left, and that leads to information
on manuals and on innovation. For the book, the papers on the meetings
of Council at ministerial level are relevant. The most recent as of 2009 is
www.oecd.org/mcm2009. Click on 'Background Reading' and the Interim
Report on the Innovation Strategy can be found, along with many other
documents of interest.

OST

The Observatoire des Sciences et des Techniques site (OST) is: www. obs-ost.fr/en.html. The OST in Paris produces indicators and reports of science, technology and innovation activities in various countries.

Statistics Canada

The URL to access the website is: www.statcan.gc.ca. On the site, there can be found survey questionnaires in English and in French, along with an explanation of the survey methodology. The route to the innovation surveys is provided in Appendix B. In addition to information on surveys, the site provides access to publications and analytical work. Some of this is discussed in Appendix B.

RICYT

The Network on science and Technology Indicators – Ibera-American and Inter-American (RICYT) is the forum for innovation indicators in Latin America and the Caribbean. The website is: www.ricyt.org and the Bogota Manual is available at www.ricyt.org/interior/difusion/pubs/ bogota/bogota_eng.pdf. The site provides a link to the Spanish literature.

Appendix B: Examples of research projects

INTRODUCTION

The purpose of this appendix is to provide examples of analytical activities related to innovation which can be undertaken by those with access to the data or those able to commission such work as part of policy analysis. The examples are drawn from work that has been done over the years at Statistics Canada, and in other institutions, and could easily be replicated in other statistical offices. Given the history of the subject, research and development (R&D) statistics are over-represented, but formal knowledge creation remains an important innovation activity which may lead to innovation.

DATA PROJECTS

Propensity to Innovate Compared with the Propensity to do R&D

The higher propensity to innovate, compared with the propensity to do R&D, in a population of firms has already been discussed in the text. The propensity to innovate is measured in the Community Innovation Survey (CIS) or CIS-like surveys. There are three ways to measure the propensity to do R&D. The R&D propensity reported in Uhrbach (2009) is based on responses to the generic CIS question 5.1 and does not distinguish between continuous and occasional. The results, presented in Table 4.1 in Chapter 4, are for a three-year period and may have a large component of occasional R&D performers. An alternative is to use the CIS question 5.2 which seeks R&D expenditure information for one year only. To get the propensity, it would be a matter of taking the counts of firms that responded, rather than the expenditure data, and then producing the population estimate for the percentage of firms that do R&D. The third method is to use the R&D survey of the country which will follow Frascati guidelines and would be expected to produce a smaller estimate, for the same set of population restrictions, than the innovation survey.

The reason for having a good estimate of the difference between the propensity to innovate and to do R&D is to understand how large is the population of firms that innovate but do no R&D. Innovation is about converting knowledge to value and the policy intervention to support non-R&D performing firms is different from that for R&D performers. It is more a matter of facilitating knowledge transfer than just applying for an R&D tax credit.

Geographical Estimates of Innovation Activity

Geography matters. States or provinces have different histories, cultures and industrial structures and their governments want to see innovation estimates for their own regions. This has implications for survey cost, respondent burden, and for survey methodology. The cost and burden issues are straightforward. The smaller the region for which statistics are to be produced, the larger the sample; and the larger the sample, the greater is the cost to the survey organization and the burden on the population. The reader should keep in mind that responding to a survey is a tax on the resources of the firm, especially when it is compulsory. The survey methodology is another matter.

To produce regional statistics, the unit being surveyed must have a location for industrial activities being performed. In North America, this is the establishment (Statistics Canada 2007), although an establishment can have more than one location. Being able to survey at the establishment level assumes a business register that contains characteristics of the firm and of its establishments with sufficient detail to support the drawing of a sample. Almost all small and medium-sized enterprises (SMEs) have one location and one industry classification. Larger firms can have more than one establishment, with different establishments classified to different industries, and being in different locations. These issues are discussed in the introduction to the North American Industry Classification System 2007 (Statistics Canada 2007). The survey methodology for the Statistics Canada Innovation Survey 2005 is described on the Statistics Canada website.[1]

Some Questions about R&D

R&D is an innovation activity and one that attracts a lot of attention in the press and policy intervention by governments. The innovation measurement practitioner, or policy analyst, should know about how R&D is distributed in their country. If they cannot get answers to the questions that follow, they should ask why not.

Table B.1   Number of R&D performers and R&D expenditures by
performing company revenue size, 2005

Revenue size	Number	% of total	Millions of dollars	% of total
Non-commercial firms	19	0.1	186	1.2
Less than 1 million	7303	38.3	1060	6.7
1.0–9.9 million	8153	42.7	2397	15.2
10.0–49.4 million	2463	12.9	1775	11.2
50.0–99.9 million	459	2.4	1038	6.6
100.0–399.9 million	432	2.3	2386	15.1
400.0 million or more	258	1.4	6949	44.0
Total	19087	100.0	15791	100.0

Source: Statistics Canada (2009).

Distribution of R&D performers
The distribution in Table B.1 shows that 38 per cent of R&D performers made less than $1 million in 2005 and they accounted for 7 per cent of the total intramural R&D expenditure. This is contrasted with firms that make $10 million or more, which account for 19 per cent of performers, but perform 77 per cent of the value of the R&D. The implications of such a distribution are that small performers of R&D may need different interventions from those needed to support R&D in large firms.

   In Canada the Scientific Research and Experimental Development (SR&ED) programme provides a refundable tax credit of 35 per cent to R&D performers that are small Canadian firms and a deductible tax credit of 20 per cent to all the rest. That is an example of a differentiated intervention to support R&D. After the dot com collapse in 2000 arguments were put that the tax credit should be refundable for all firms. The motivation for this was that some large firms did not have taxable revenue for some years and could not use the tax credit.

   A related observation is that R&D performance is concentrated and the top 75 performers in Canada account for about 50 per cent of the R&D performed (Statistics Canada 2009). This suggests that talking to the top 75 firms might be a first step in addressing R&D policy questions.

How persistent are R&D performers?
In a study published in 2006 (Schellings and Gault 2006), a panel of R&D performers was constructed from the database used to publish annual cross-sectional data. The period of the study was nine years and there

were about 31 000 firms in the panel. Over that period, about 10 000 R&D performers were identified each year.

The R&D performers were classified by performance size and the striking statistic was that 64 per cent of the firms in the panel performed less than $100 000 of R&D, not enough to pay a full-time engineer. The second statistic of interest was that 25 per cent of the panel performed less than $100 000 and were present for only one year and 41 per cent were present for one or two years. This suggests that the occasional R&D performer is dominant in Canadian statistics and this has implications for policies to promote R&D performance as part of the innovation process.

The same paper looked at the survival as an R&D performer for firms that entered the panel, according to their performance size. For those firms performing less than $100 000, most had vanished by the end of nine years but a small percentage had become large R&D performers. Were such a study ever repeated, it could be complemented by case studies to understand the factors that supported the growth of R&D performance in those cases where it took place.

R&D intensity

R&D intensity is the ratio of the value of R&D performed by a firm to its revenue. This ratio varies with the industry in which the R&D is performed, and it has been suggested that it is related to the lifetime of the products produced by the industry. Size is an important factor in its interpretation as can be seen from Table B.2.

R&D performers in the smallest revenue size class, especially when they are start-up firms, will have a high ratio. This falls as revenue size

Table B.2   Current intramural R&D expenditures as a percentage of performing company revenues, by performing company revenue size, 2005

	2003	2004	2005	2006
Revenue size	1.8	1.8	1.7	1.7
Less than 1 million	45.4	48.5	38.7	38.1
1.0–9.9 million	8.8	7.5	7.7	7.2
10.0–49.9 million	3.5	3.6	3.2	3.6
50.0–99.9 million	2.7	2.7	3.1	2.8
100.0–399.9 million	2.6	2.9	2.7	2.5
400.0 million or more	1.1	1.0	1.0	1.0

Source: Statistics Canada (2009).

increases, and in Canada the mid-range firms attract takeover bids from foreign firms. As the size goes up, the ratio goes down.

The statistic is used for the comparison of the behaviour of foreign and domestically controlled firms. There is a case for doing the comparison by size class as the size distribution of foreign-controlled firms is different from that of the domestic ones.

How do Firms Grow?

The paper by Bordt et al. (2004) goes beyond paper or electronic survey to include interviews and analysis of the findings. The paper looks at growth factors and these should be reviewed with the question in mind of how public policy could promote growth. The finding, highlighted in the text, is that there are some firms where growth is not an objective. There are many reasons for this, but they should be understood before launching a growth or 'gazelle' promotion.

The URL for the study is: www.statcan.gc.ca/pub /88f0006x/ 88f0006x2004021-eng.pdf. If the subject is of interest, on the site, click on 'Publications' and search by name of any of the authors.

Linkages

An early experiment with bibliometrics gave rise to a paper which shows the linkage between institutions and regions based on a bibliometric database that could associate a geographical location with an author name. This showed vividly how academic collaboration took place in Canada, and it could be replicated elsewhere. The experiment was so successful that it gave rise to the Observatoire des sciences et des technologies which continues to produce bibliometric products. The website is www.ost.uqam.ca/Observatoire/tabid/56/language/en-US/Default. aspx.

The paper that gave rise to it is Davignon et al. (1998). The URL is www.statcan.gc.ca/pub/88f0006x/88f0006x1998010-eng.pdf.

Other Topics

The reader is encouraged to search by topic to find papers on the site. Searching for Anderson, April, Bordt, Earl, Lonmo, Rose, Schaan or Sciadas will provide a number of papers that could be reproduced with current data.

NOTE

1. On the site, www.statcan.gc.ca, click on 'Science and technology' in the 'Browse by box, then click on 'Definitions, data source and methods', which is on the left of the page. While navigating to the survey methodology, pause to note the various classification systems, including the North American Industry Classification System (NAICS). The material on surveys and on classifications should equip the reader with enough knowledge to ask informed questions of their own statistical office or research institute.

References

AAAS (2009), 'Special Section: Complex Systems and Networks', *Science*, **325**, 405–32.

Anlló, Guillermo (2006), 'An Overview of Latin American Innovation Surveys', in William Blankley, Mario Scerri, Neo Molotja and Imraan Saloojee (eds), *Measuring Innovation in OECD and Non-OECD Countries*, Cape Town: HSRC Press, pp. 141–62.

Antal, Berthoin Ariane, Meinolf Dierkes, John Child and Ikujiro Nonaka (2001), 'Organizational Learning and Knowledge, Reflections on the Dynamics of the Field and Challenges for the Future', in M. Dierkes, A. Bethoin Antal, J. Child and I. Nonaka (eds), *Handbook of Organizational Learning and Knowledge*, Oxford: Oxford University Press, pp. 921–39.

Arundel, Anthony (2007), 'Innovation Survey Indicators: What Impact on Innovation Policy?' in OECD, *Science, Technology and Innovation Indicators in a Changing World: Responding to Policy Needs*, Paris: OECD, pp. 49–64.

Arundel, Anthony, Catalina Bordoy and Minna Kanerva (2008a), *Neglected Innovators: How Do Innovative Firms that Do not Perform R&D Innovate? Results of an Analysis of the Innobarometer 2007 Survey 215*, INNO-Metrics Thematic Paper, 31 March.

Arundel, Anthony, Cati Bordoy, Pierre Mohnen and Keith Smith (2008b), 'Innovation Surveys and Policy: Lessons from the CIS', in Claire Nauwelaers and René Wintjes (eds), *Innovation Policy in Europe, Measurement and Strategy*, Cheltenham, UK and Northampton, MA, USA: Edward Elgar, pp. 3–28.

Arundel, Anthony and Viki Sonntag (1999), *Patterns of Advanced Manufacturing Technology (AMT) Use in Canadian Manufacturing: 1998 AMT Survey Results*, Catalogue no. 88F0017MIE, no. 12, Ottawa: Statistics Canada.

Atkinson, Robert D. (2004), *The Past and Future of America's Economy: Long Waves of Innovation that Power Cycles of Growth*, Cheltenham, UK and Northampton, MA, USA: Edward Elgar.

Atkinson, Robert and Howard Wial (2008), 'Creating a National Innovation Foundation', *Issues in Science and Technology*, Fall, 75–84.

Aubert, Jean-Eric (2004), *Promoting Innovation in Developing Countries, A Conceptual Framework*, Policy Research Working Paper 0-3097, Washington, DC: World Bank Institute.

Aubert, Jean-Eric (2006), 'Innovation Systems in Emerging and Developing Economies', in William Blankley, Mario Scerri, Neo Molotja and Imraan Saloojee (eds), *Measuring Innovation in OECD and Non-OECD Countries*, Cape Town: HSRC Press, pp. 141–62.

Auriol, Laudeline (2007), 'The International Mobility of Doctorate Holders: First Results and Methodological Advances', in OECD, *Science, Technology and Innovation Indicators in a Changing World: Responding to Policy Needs*, Paris: OECD, pp. 193–212.

Ayres, R.U. (1978), *Resource, Environment and Economic: Applications of the Material/Energy Balance Principle*, New York: J. Wiley & Sons.

Baczko, Tadeusz (ed.) (2009), *The Future of Science and Technology and Innovation Indicators and the Challenges Implied*, Warsaw: Institute of Economics, Polish Academy of Sciences.

Barber, Douglas H. and Jeffrey Crelinsten (2009), *Understanding the Disappearance of Early-Stage and Start-Up R&D Performing Firms*, Toronto: Impact Group.

BEA (2007), *Research and Development Satellite Account, 2007 Satellite Account Underscores Importance of R&D*, BEA-07-48, Washington, DC: BEA.

Beattie, Alan (2009), *False Economy: A Surprising Economic History of the World*, Toronto: Viking Canada.

Bernstein, Alan, Vern Hicks, Peggy Boorbey, Terry Campbell, Laura McAuley and Ian D. Graham (2007), 'A Framework to Measure the Impacts of Investments in Health Research', in OECD, *Science, Technology and Innovation Indicators in a Changing World: Responding to Policy Needs*, Paris: OECD, pp. 231–50.

Blankley, William, Mario Scerri, Neo Molotja and Imraan Saloojee (eds) (2006), *Measuring Innovation in OECD and Non-OECD Countries*, Cape Town: HSRC Press.

Blind, Knut, Jakob Edler, Luke Georghiou, Elvira Uyarra, Deborah Cox, John Rigby and Yanuar Nugroho (2009), *Monitoring and Evaluation Methodology for the EU Lead Market Initiative, A Concept Development, Final Report*, Manchester: Manchester Institute of Innovation Research, University of Manchester.

Block, Carter and Vladimir López-Bassols (2009), 'Innovation Indicators', in OECD, *Innovation in Firms: A Microeconomic Perspective*, Paris: OECD, pp. 21–68.

BMBF (2006), *The High-Tech Strategy for Germany*, Bonn and Berlin: BMBF.

BMBF (2008a), *Bundesbericht Forschung und Innovation 2008*, Bonn and Berlin: BMBF.

BMBF (2008b), *Strengthening Germany's Role in the Global Knowledge Society*, Bonn and Berlin: BMBF.

BMBF (2008c), *10 Thesen für ein starkes Wissenschaftssystem im weltweiten Wettbewerb, Demands on Research Landscapes under Changing Framework Conditions, Memorandum*, Bonn and Berlin: BMBF.

Boden, Mark and Ian Miles (eds) (2000), *Services and the Knowledge-Based Economy*, London: Continuum.

Bordt, Michael, Louise Earl, Charlene Lonmo and Robert Joseph (2004), *Characteristics of Firms that Grow from Small to Medium Size: Growth Factors – Interviews and Measurability*, Catalogue 88F0006XIE2004021, Ottawa: Statistics Canada.

Bordt, Michael, Julio Miguel Rosa and Johanne Boivin (2007), 'Science, Technology and Innovation for Sustainable Development: Towards a Conceptual Framework', in OECD, *Science, Technology and Innovation Indicators in a Changing World: Responding to Policy Needs*, Paris: OECD, pp. 251–68.

Brown, Lawrence D. Thomas J. Plewes and Marisa A. Gerstein (eds) (2005), *Measuring Research and Development Expenditures in the US Economy*, Panel on Research and Development Statistics at the National Science Foundation, Committee on National Statistics, Division of Behavioral and Social Sciences and Education. Washington, DC: National Academies Press.

Business Week (2008), 'Can America Invent its Way Back', 11 September.

Carlsson, Bo (ed.) (1997), *Technological Systems and Industrial Dynamics*, Boston, MA: Kluwer Academic Publishers.

CCA (2009a), *Report in Focus, Innovation and Business Strategy: Why Canada Falls Short*, Ottawa: Council of the Canadian Academies, www.scienceadvice.ca/innovation.html.

CCA (2009b), *Innovation and Business Strategy: Why Canada Falls Short*, Ottawa: Council of the Canadian Academies, www.scienceadvice.ca/innovation.html.

CEC (2005), *Proposal for a Decision of the European Parliament and the Council Establishing a Competitiveness and Innovation Framework Programme (2007–2013)*, COM (2005) 121 Final, Brussels: Commission of the European Communities.

CEC (2006a), *Creating an Innovative Europe: Report of the Independent Expert Group on R&D and Innovation following the Hampton Court Summit*, chaired by Mr Esko Aho, Brussels: Commission of the European Communities.

CEC (2006b), *Putting Knowledge into Practice: A Broad-Based Innovation*

Strategy for the EU, COM (2006) 502 final, Brussels: Commission of the European Communities.

CEC (2007a), *Mid-term Review of Industrial Policy: A Contribution to the EU's Growth and Jobs Strategy*, COM (2007) 374, Brussels: Commission of the European Communities.

CEC (2007b), *A Lead Market Initiative for Europe*, SEC(2007), 1729,1730, COM(2007) 860 final, Brussels: Commission of the European Communities.

CEC (2007c), *Towards a European Strategy in Support of Innovation in Services: Challenges and Key Issues for Future Actions*, SEC (2007) 1059, Brussels: Commission of the European Communities.

CEC (2007d), *Removing Obstacles to Cross-Border Investments by Venture Capital Funds*, Brussels: Commission of the European Communities.

CEC (2008a), *Towards World-Class Clusters in the European Union: Implementing the Broad-Based Innovation Strategy*, SEC (2008) 2673, Brussels: Commission of the European Communities.

CEC (2008b), *Annex to the Communication from the Commission, 'Towards World-Class Clusters in the European Union: Implementing the Broad-Based Innovation Strategy'*, COM (2008) 652 final of 17.10.2008, Brussels: Commission of the European Communities.

CEC (2008c), *An Industrial Property Rights Strategy for Europe*, COM (2008) 465 final, Brussels: Commission of the European Communities.

CEC (2008d), *Towards an Increased Contribution from Standardization to Innovation in Europe*, COM (2008) 133 final, Brussels: Commission of the European Communities.

CEC (2008e), *Action Plan on the Sustainable Consumption and Production and Sustainable Industrial Policy*, COM (2008) 397 final, Brussels: Commission of the European Communities.

CEC (2009a), *Design as a Driver of User-Centered Innovation*, Commission Staff Working Document SEC (2009) 501 final, Brussels: Commission of the European Communities.

CEC (2009b), *Mainstreaming Sustainable Development into EU Policies: 2009 Review of the European Union Strategy for Sustainable Development*, COM (2009) 400 final, Brussels: Commission of the European Communities.

CEC (2009c), *Reviewing Community Innovation Policy in a Changing World*, COM(2009) 442 final, Brussels: Commission of the European Communities.

CEC IMF OECD UN World Bank (1994), *System of National Accounts 1993*, New York: United Nations.

CES (2008a), *Development of a Handbook on Deriving Capital Measures of Intellectual Property Products*, Conference of European Statisticians,

Joint UNECE/Eurostat/OECD Meeting on National Accounts, April, ECE/CES/GE.20/2008/5, Geneva: CES.

CES (2008b), *1993 SNA Update Issues, Research and Development, ECE/CES/GE.20/2008/13*, Geneva: UN Economic and Social Council, Economic Commission for Europe, Council of European Statisticians.

Chesbrough, Henry (2003), *Open Innovation: The New Imperative for Creating and Profiting from Technology*, Boston, MA: Harvard University Press.

Christensen, C.M. (1997), *The Innovators Dilemma: When New Technologies Cause Great Firms to Fail*, Boston, MA: Harvard University Press.

Christensen, Clayton M. (2008), 'Foreword: Reflections on Disruption', in Scott D. Anthony, Mark W. Johnson, Joseph V. Sinfield and Elizabeth J. Altman (eds), *The Innovator's Guide to Growth: Putting Disruptive Innovation to Work*, Boston, MA: Harvard Business Press, pp. vii–xiv.

Collier, Paul (2007), *The Bottom Billion, Why the Poorest Countries Are Failing and What Can Be Done About It*, Oxford: Oxford University Press.

Collier, Paul (2008), 'The Politics of Hunger: How Illusion and Greed Fan the Food Crisis', *Foreign Affairs*, **87**(6), 67–79.

Conference Board (2008), *Workshop on Developing a New National Research Data Infrastructure for the Study of Organizations and Innovation: Workshop Report*, Washington, DC: Conference Board.

Council of Science and Technology, Japan (2008), *Toward the Reinforcement of Science and Technology Diplomacy, Preliminary Translation*, 19 May 2008, Tokyo: Council of Science and Technology.

Dahlman, Carl J., Jorma Routti and Pekka Ylä-Anttila (eds) (2006), *Finland as a Knowledge Economy: Elements of Success and Lessons Learned*, Washington, DC: World Bank.

David, P.A. (1993), 'Knowledge, Property and the System Dynamics of Technological Change', in World Bank (ed.), *Proceedings of the World Bank Annual Conference on Development Economics 1992*, Washington, DC: World Bank, pp. 215–48.

David, P.A. and D. Foray (1995), 'Accessing and Expanding the Science and Technology Knowledge Base', *STI Review*, **16**, 13–68.

Davigon, Louis, Yves Gingras and Benoit Godin (1998), *Knowledge Flows in Canada as Measured by Bibliometrics*, Ottawa: Statistics Canada.

de Jong, Jeroen P.J. and Eric von Hippel (2009), *User Innovation in Dutch High-Tech SMEs: Frequency, Nature and Transfer to Producers*, MIT Sloan School of Management Working Paper no. 4724–09, Cambridge, MA: MIT.

de la Mothe, John and Dominique Foray (eds) (2001), *Knowledge*

Management in the Innovation Process, Boston, MA: Kluwer Academic Publishers.

DFID (2008), *Research Strategy 2008–2013*, London: DFID.

Diamond, Jared (1997), *Guns, Germs, and Steel*, New York: W.W. Norton.

Dierkes, M. (2001), 'Visions, Technology, and Organizational Knowledge: An Analysis of the Interplay between Enabling Factors and Triggers of Knowledge Generation', in John de la Mothe and Dominique Foray (eds), *Knowledge Management in the Innovation Process*, Boston, MA: Kluwer Academic Publishers, pp. 11–42.

Dierkes, Meinolf, Ariane Berthoin Antal, John Child and Ikujiro Nonaka (eds) (2001a), *Handbook of Organizational Learning and Knowledge*, Oxford: Oxford University Press.

Dierkes, Meinolf, Marcus Alexis, Ariane Berthoin Antal, Bo Hedberg, Peter Pawlowsky, John Stopford and Anne Vonderstein (eds) (2001b), *The Annotated Bibliography of Organizational Learning and Knowledge Creation*, 2nd edn, Berlin: Wissenschaftszentrum Berlin für Sozialforschung (WZB).

DIUS (2008), *Innovation Nation*, Department for Innovation, Universities and Skills, Norwich: HMSO.

Dodgson, Mark and Roy Rothwell (eds) (1994), *The Handbook of Industrial Innovation*, Aldershot, UK and Brookfield, VT, USA: Edward Elgar.

Ducharme, L.M. and Fred Gault (1992), 'Surveys of Manufacturing Technology', *Science and Public Policy*, **19**, 393–9.

Dyson, Freeman (2007), 'Our Biotech Future', *New York Review of Books*, **54**(12), 4–8.

Earl, Louise (2002), 'Putting Your Money where Your Mouth Is: Using Knowledge Management Practices to Design a Knowledge Management Survey', *Innovation Analysis Bulletin*, **4**(1), Catalogue 88-003-XIE, page 11, Ottawa: Statistics Canada.

Earl, Louise (2003), 'Are we Managing our Knowledge? The Canadian Experience', in OECD, *Measuring Knowledge Management in the Business Sector: First Steps*, Paris: OECD, pp. 55–87.

Earl, Louise and Fred Gault (2003), 'Knowledge Management: Size Matters', in OECD, *Measuring Knowledge Management in the Business Sector: First Steps*, Paris: OECD, pp. 169–86.

Earl, Louise and Fred Gault (2006), *National Innovation, Indicators and Policy*, Cheltenham, UK and Northampton, MA, USA: Edward Elgar.

Edler, Jakob and Luke Georghiou (2007), 'Public Procurement and Innovation – Resurrecting the Demand Side', *Research Policy*, **36**, 949–63.

Edquist, Charles (1997), *Systems of Innovation, Technologies, Institutions and Organizations*, London: Pinter.

Edquist, Charles (2004), 'Systems of Innovation: A Critical Review of the State of the Art', in J. Fagerberg, D. Mowery and R. Nelson (eds), *The Oxford Handbook of Innovation*, Oxford: Oxford University Press.

Ellis, Simon (2008), 'The Current State of International Science Statistics in Africa', *African Statistical Journal*, **6**, 177–89.

Executive Office of the President (2009), *A Strategy for American Innovation: Driving Towards Sustainable Growth and Quality Jobs*, Washington, DC: Executive Office of the President/NEC/OSTP.

Fabling, Richard (2007), 'How Innovative are New Zealand Firms? Quantifying and Relating Organizational and Marketing Innovation to Traditional Science and Technology Indicators', in OECD, *Science, Technology and Innovation Indicators in a Changing World: Responding to Policy Needs*, Paris: OECD, pp. 139–70.

Fagerberg, Jan, David C. Mowery and Richard Nelson (eds) (2004), *The Oxford Handbook of Innovation*, Oxford: Oxford University Press.

Farmer, Doyne J., William Brian Arthur, Jessika Trancik, Douglas H. Erwin and Walter W. Powell (2007), *Modelling the Dynamics of Technological Evolution, NSF Proposal*, Washington, DC: NSF, http://www.nsf.gov/sbe/scisip/scisipnews1.pdf.

Florida, Richard (1998), 'Calibrating the Learning Region', in John de la Mothe and Gilles Paquet (eds), *Local and Regional Systems of Innovation*, Boston, MA: Kluwer Academic Publishers.

Florida, Richard (2002), *The Rise of the Creative Class: And how It's Transforming Work, Leisure, Community and Everyday Life*, Philadelphia, PA: Basic Books.

Foray, D. (2004), *The Economics of Knowledge*, Cambridge, MA: MIT Press.

Foray, Dominique (2007), 'Enriching the Indicator Base for the Economics of Knowledge', in OECD, *Science, Technology and Innovation Indicators in a Changing World: Responding to Policy Needs*, Paris: OECD, pp. 87–100.

Forrester, J.W. (1971), *World Dynamics*, Cambridge, MA: Wright-Allen Press.

Forrester, J.W. (1982), 'Global Modelling Revisited', *Futures*, **14**, 95–110.

Franke, N. and S. Shah (2003), 'How Communities Support Innovative Activities: An Exploration of Assistance and Sharing Among End-Users', *Research Policy*, **32**(1), 157–78.

Franke, N. and Eric von Hippel (2003), 'Satisfying Heterogeneous User Needs via Innovation Toolkits: The Case of Apache Security Software', *Research Policy*, **32**(7), 1199–1215.

Freeman, C. (1987), *Technology Policy and Economics Performance: Lessons from Japan*, London: Pinter.

Freeman, Chris and Luc Soete (2007), 'Developing Science and Technology and Innovation Indicators: The Twenty-First Century Challenges', in OECD, *Science, Technology and Innovation Indicators in a Changing World: Responding to Policy Needs*, Paris: OECD, pp. 271–84.

Friedman, Thomas L. (2006), *The World is Flat: A Brief History of the Twenty-First Century*, 2nd edn, New York: Farrar, Straus & Giroux.

Friedman, Thomas L. (2008), *Hot, Flat and Crowded: Why we Need a Green Revolution – and How it Can Renew America*, New York: Farrar, Straus & Giroux.

Gadrey, Jean and Faïz Gallouj (eds) (2002), *Productivity, Innovation and Knowledge in Services, New Economic and Socio-Economic Approaches*, Cheltenham, UK and Northampton, MA, USA: Edward Elgar.

Galindo-Rueda, Fernando (2007), 'Developing an R&D Satellite Account for the UK: A Preliminary Analysis', *Economic and Labour Market Review*, **1**, 18–29.

Gallouj, Faïz (2002), *Innovation in the Service Economy: The New Wealth of Nations*, Cheltenham, UK and Northampton, MA, USA: Edward Elgar.

Gault, F.D. (1998), 'The Federal Strategy for Science and Technology in Canada and Statistical Measurement', in A.M. Herzberg and I. Krupka (eds), *Statistics, Science and Public Policy*, Kingston Ont.: Queen's, pp. 181–8.

Gault, Fred (2004), 'Developing and Using Indicators of Science and Technology Activity: Canadian Experience in an International Context', in *Proceedings of the Seventh Forum on International Science and Technology Indicators, October 21–23, 2004, Beijing, China*, pp. 30–45 (in Chinese).

Gault, Fred (2007a), 'Science, Technology and Innovation Indicators: The Context of Change', in OECD, *Science, Technology and Innovation Indicators in a Changing World: Responding to Policy Needs*, Paris: OECD, pp. 9–23.

Gault, Fred (2007b), 'Assessing International S&T Co-operation for Sustainable Development: Towards Evidence-based Policy', in OECD, *Integrating Science and Technology into Development Policies: An International Perspective*, Paris: OECD.

Gault, Fred (2008a), 'Indicadores de ciencia, tecnología e innovación: próximos pasos', *en Indicadores de Ciencia y Tecnología en Iberoamérica – Agenda 2008*, Buenos Aires: RICYT, pp. 37–49.

Gault, Fred (2008b), 'Science, Technology and Innovation Indicators: Opportunities for Africa', *African Statistical Journal*, **6** (May), 141–62.

Gault, Fred (2009), 'The OECD Innovation Strategy: Delivering Value', *Foresight*, **1**(9), 16–28 (in Russian).

Gault, F.D., K.E. Hamilton, R.B. Hoffman and B.C. McInnis (1987), 'The Design Approach to Socio-Economic Modelling', *Futures*, **19**, 3–25.

Gault, F.D., R.B. Hoffman and B.C. McInnis (1985), 'The Path to Process Data', *Futures*, **17**, 509–27.

Gault, Fred and Susanne Huttner (2008), 'Commentary: A Cat's Cradle for Policy', *Nature*, **455**, 462–3.

Gault, F.D. and S. McDaniel (2002), 'Continuities and Transformations: Challenges to Capturing Information about the "Information Society"', *First Monday*, **7**(2), 1–13.

Gault, Fred and William Pattinson (1994), *Model Surveys of Service Industries: The Need to Measure Innovation*, Voorburg Conference Paper, Sydney, Australia.

Gault, Fred and William Pattinson (1995), *Innovation in Service Industries: The Measurement Issues*, Voorburg Conference Paper, Voorburg, The Netherlands.

Gault, F.D., B.J. Read, P.R. Stevens and A. Rittenberg (1979), 'The Use of Database Management Systems in Particle Physics', in B. Dreyfus (ed.), *Proceedings of the Sixth International CODATA Conference (Sicily 1978)*, Oxford: Pergamon Press, Oxford, pp. 167–9.

Gault, Fred and Eric von Hippel (2009), *The Prevalence of User Innovation and Free Innovation Transfers: Implications for Statistical Indicators and Innovation Policy*, MIT Sloan School of Management Working Paper no. 4722-09, Cambridge, MA: MIT.

Gault, Fred and Gang Zang (2009), 'The View from the Workshop', in *Innovation for Development: Converting Knowledge to Value, Summary Report, Paris, 28 to 30 January 2009*, Paris: UNESCO. pp. 3–6.

Georghiou, Luke, Jennifer Cassingena Harper, Michael Keenan, Ian Miles and Rafael Popper (eds) (2008), *The Handbook of Technology Foresight*, Cheltenham, UK and Northampton, MA, USA: Edward Elgar.

Government of Canada (2007), *Mobilizing Science and Technology to Canada's Advantage*, Ottawa: Government of Canada.

Government of Canada (2009), *Canada's Science, Technology and Innovation System: State of the Nation 2008*, Ottawa: Government of Canada.

Government of Denmark (2006), *Progress, Innovation and Cohesion: Strategy for Denmark in the Global Economy*, Copenhagen: Government of Denmark.

Government of Finland (2008a), *Steering Group Proposal for a National Innovation Strategy, 'Finland's National Innovation Strategy'*, Helsinki: Government of Finland.

Government of Finland (2008b), *Government's Communication on Finland's National Innovation Strategy to Parliament*, Helsinki: Government of Finland.

Government of Sweden, Ministry of Industry, Employment and Communications (2004), *Innovative Sweden: A Strategy for Growth through Renewal*, Stockholm: Ministry of Industry, Employment and Communications.

Government of the Netherlands (2008), *Long-Term Strategy: Towards an Agenda for Sustainable Growth in Productivity*, The Hague: Ministry of Economic Affairs and Ministry of Education, Culture and Science.

Hall, B. and N. Rosenberg (eds) (2010), *Handbook of Innovation*, Amsterdam, The Netherlands and New York, USA: Elsevier.

Harayama, Yuko (2007), *International Cooperation in Japanese Science and Technology Policy*, http://ec.europa.eu/research/iscp/index.cfm?lg=en&pg=wkshp_25-26_09_2007.

Hawkins, Richard W., Cooper H. Langford and Kiranpal S. Sidhu (2007), 'University Research in an "Innovation Society"', in OECD, *Science, Technology and Innovation Indicators in a Changing World: Responding to Policy Needs*, Paris: OECD, pp. 171–92.

Henkel, Joachim and Stephanie Pangerl (2008), *Defensive Publishing: An Empirical Study*, Working Paper, Munich: Technical University of Munich.

Herstatt, C. and Eric von Hippel (1992), 'From Experience Developing New Product Concepts via the Lead User Method', *Journal of Product Innovation Management*, **9**(3), 213–22.

Hienerth, Christoph (2006), 'The Commercialization of User Innovations: The Development of the Rodeo Kayak Industry', *R&D Management*, **36**, 273–94.

HM Government (2009), *New Industry New Jobs*, London: HM Government.

HM Treasury / BERR (2008), *Enterprise: Unlocking the UK's Talent*, London: HM Treasury.

HRDC (2002), *Knowledge Matters: Skills and Learning for Canadians*, Ottawa: HRDC.

Industry Canada (1996), *Science and Technology for the New Century: A Federal Strategy*, Ottawa: Government of Canada, Supply and Services Canada.

Industry Canada (2001), *Achieving Excellence: Investing in People, Knowledge and Opportunity: Canada's Innovation Strategy – Main Report*, Ottawa: Government of Canada.

Innovation Platform (2009), *Stronger After the Storm, Investing in*

People and Knowledge to Emerge from the Crisis Stronger, The Hague: Innovation Platform.

Jaffe, Adam B., Josh Lerner and Scott Stern (2006), *Innovation Policy and the Economy*, Vol. 6, Cambridge: MIT Press.

Janz, N., G. Ebling, S. Gottschalk and H. Niggemann (2001), 'The Mannheim Innovation Panels (MIP and MIP-S) of the Centre for European Economic Research (ZEW)', *Journal of Applied Social Science Studies*, **121**(1), 123–9.

Kahn, Michael (2008), 'Africa's Plan of Action for Science and Technology and Indicators: South African Experience', *African Statistical Journal*, **6**(May), 163–76.

Kapstein, Ethan B. (2009), 'Africa's Capitalist Revolution, Preserving Growth in a Time of Crisis', *Foreign Affairs*, **88**(4), 119–28.

Kremp, E. and J. Mairesse (2002), *Le 4 Pages des statistiques industrielles*, No. 169, December, Paris: SESSI.

Kuznetsov, Yevgeny (2006), *Diaspora Networks and the International Migration of Skills: How Countries Can Draw on their Talent Abroad*, Washington, DC: World Bank Institute Development Studies.

Licht, Georg (2008), 'Nachgefragt: Innovationsverhalten von KMU – Steuerpolitik ist Innovationspolitik', Mannheim: *ZEWnews* Juli/August, 3.

List, Friedrich (1841/1959), *Das Nationale System der Politischen Oekonomie*, Basel: Kyklos-Verlag.

List, Friedrich (1909), *The National System of Political Economy*, J. Shield Nicholson (ed.), Sampson S. Lloyd (trans.), London: Longmans, Green & Co. Library of Economics and Liberty. Available from http://www.econlib.org/library/YPDBooks/List/1stNPEO.html, accessed 13 July 2009.

LO (2008), *Employee-Driven Innovation*, Copenhagen: LO, Danish Confederation of Trade Unions.

Lugones, Gustavo (2006), 'The Bogotá Manual: Standardising Innovation Indicators for Latin America and the Caribbean', in William Blankley, Mario Scerri, Neo Molotja and Imraan Saloojee (eds), *Measuring Innovation in OECD and Non-OECD Countries*, Cape Town: HSRC Press, pp. 163–81.

Lundvall, B.-Å. (ed.) (1992), *National Innovation Systems: Towards a Theory of Innovation and Interactive Learning*, London: Pinter.

Lundvall, B.-Å (2007), *Innovation System Research, Where it Came From and Where it Might Go*, Globelics Working Paper Series, No. 2007–01, Aalborg, Denmark: Globelics.

Lundvall, B.-Å. and B. Johnson (1994), 'The Learning Economy', *Journal of Industry Studies*, **1**(2), pp. 23–4.

Lundvall, B.-Å., K.J. Joseph, Cristina Chaminade and Jan Vang (2009), *Handbook of Innovation Systems and Developing Countries: Building Domestic Capabilities in a Global Setting*, Cheltenham, UK and Northampton, MA, USA: Edward Elgar.

Lüthje, C. (2003), 'Customers as Co-Inventors: An Empirical Analysis of the Antecedents of Customer-Driven Innovations in the Field of Medical Equipment', in *Proceedings of the 32th EMAC Conference*, Glasgow.

Lüthje, C. (2004), 'Characteristics of Innovating Users in a Consumer Goods Field: An Empirical Study of Sport-Related Product Consumers', *Technovation*, **24**(9), 683–95.

Lüthje, C., C. Herstatt and E. von Hippel (2002), *The Dominant Role of Local Information in User Innovation: The Case of Mountain Biking*, Working Paper, MIT Sloan School of Management.

Macher, Jeffery T. and David C. Mowery (eds) (2008), *Innovation in Global Industries: US Firms Competing in a New World*, Washington, DC: National Academies Press.

Mansfield, Edwin (1968), *The Economics of Technological Change*, New York: W.W. Norton & Company.

Marburger, John (2005), 'Wanted: Better Benchmarks', *Science*, **308**(5725), 1087.

Marburger, John (2007), 'The Science of Science and Innovation Policy', in OECD, *Science, Technology and Innovation Indicators in a Changing World: Responding to Policy Needs*, Paris: OECD, pp. 27–32.

McDaniel, Susan (2006), 'Innovation in Human/Social Guise', in Louise Earl and Fred Gault (eds), *National Innovation, Indicators and Policy*, Cheltenham, UK and Northampton, MA, USA: Edward Elgar, pp. 154–64.

Meadows, Donella H., Dennis L. Meadows and Jorgen Randers (1992), *Beyond the Limits*, Post Mills, VT: Chelsea Green Publishing Company.

Meadows, Donella H., Dennis L. Meadows, Jorgen Randers and William W. Behrens III (1972), *The Limits to Growth*, New York: Universe Books.

Meadows, Donella, Jorgen Randers and Dennis Meadows (eds) (2004), *Limits to Growth: The 30-Year Update*, White River Junction, VT: Chelsea Green Publishing Company.

Metcalfe, Stanley J. and Ian Miles (eds) (2000), *Innovation Systems in the Service Economy: Measurement and Case Study Analysis*, Norwell, MA: Kluwer Academic Publishers.

Miles, Ian, Jennifer Cassingena Harper, Luke Georghiou, Michael Keenan and Rafael Popper (2008), 'The Many Faces of Foresight', in

Luke Georghiou, Jennifer Cassingena Harper, Michael Keenan, Ian Miles and Rafael Popper (eds), *The Handbook of Technology Foresight*, Cheltenham, UK and Northampton, MA, USA: Edward Elgar, pp. 3–23.

Morrison, P.D., J.H. Roberts, and E. von Hippel (2000), 'Determinants of User Innovation and Innovation Sharing in a Local Market', *Management Science*, **46**(12), 1513–27.

Mowery, David C. (ed.) (1999), *US Industry in 2000: Studies in Competitive Performance*, Washington, DC: National Academies Press.

Muldur, Ugur, Fabienne Corvers, Henri Delanghe, Jim Dratwa, Daniela Heimberger, Brian Sloan and Sandrijn Vanslembrouck (2006), *A New Deal for an Effective European Research Policy: The Design and Impacts of the 7th Framework Programme*, Dordrecht: Springer.

National Academy of Sciences / National Academy of Engineering / Institute of Medicine (2007), *Rising Above the Gathering Storm: Energizing and Employing America for a Brighter Economic Future*, Washington, DC: National Academy Press.

Nauwelaers, Claire (2009), *Policy Mixes for R&D in Europe*, Maastricht: UNU-MERIT, ec.europa.eu/research/policymix.

Nauwelaers, Claire and René Wintjes (eds) (2008), *Innovation Policy in Europe: Measurement and Strategy*, Cheltenham, UK and Northampton, MA, USA: Edward Elgar.

Nelson, Richard R. (1987), *Understanding Technical Change as an Evolutionary Process*, Amsterdam: North-Holland.

Nelson, Richard R. (1988), 'Institutions Supporting Technical Change in the United States', in Giovanni Dosi, Christopher Freeman, Richard Nelson, Gerald Silverberg and Luc Soete (eds), *Technical Change and Economic Theory*, London: Pinter, pp. 312–29.

Nelson, Richard R. (ed.) (1993), *National Systems of Innovation*, New York: Oxford University Press.

Nelson, Richard R. and Sidney G. Winter (1982), *An Evolutionary Theory of Economic Change*, Cambridge, MA: Belknap Press.

NEPAD (2003), *Declaration of the First NEPAD Ministerial Conference on Science and Technology*, 7 November, Johannesburg, South Africa, Pretoria: NEPAD.

NEPAD (2005a), *Resolutions of the Second African Ministerial Conference on Science and Technology*, 30 September, Dakar, Senegal, Pretoria: NEPAD.

NEPAD (2005b), *Inter-Governmental Committee on Science, Technology and Innovation Indicators*, Terms of Reference, Pretoria: NEPAD.

NEPAD (2005c), *Inter-governmental Committee on Science, Technology and Innovation Indicators, Terms of Reference*, Pretoria: NEPAD.

NEPAD (2006a), *African Science, Technology and Innovation Indicators (ASTII): Towards African Indicator Manuals – A Discussion Document*, www.nepadst.org/doclibrary/pdfs/iastii_jun2006.pdf.

NEPAD (2006b), *Africa's Science and Technology Consolidated Plan of Action*, Pretoria: NEPAD.

NEPAD (2006c), *Resolutions of the Sixth Meeting of the Steering Committee of the African Ministerial Council on Science and Technology (AMCOST)*, 20–21 March, Pretoria: NEPAD.

NEPAD (2006d), *Extraordinary Conference of the African Ministers Council on Science and Technology (AMCOST), November 20–24, Cairo, Egypt, Report of the Meeting of Ministers, EXT/AU/MIN/ST/ Prt.(II)*, Pretoria: NEPAD.

NEPAD (2006e), *Governing Science, Technology and Innovation in Africa, Building National and Regional Capacities to Develop and Implement Strategies and Policies: A Programme Proposal Submitted to the Swedish Agency for Research Cooperation (SAREC) of the Swedish International Development Cooperation Agency (SIDA), Stockholm, Sweden, October 2006*, Pretoria: NEPAD.

NEPAD (2007), *Decisions of the First Meeting of the African Intergovernmental Committee on Science, Technology and Innovation Indicators, September 18, 2007, Maputo, Mozambique*, Pretoria: NEPAD.

NSTC (2008), *The Science of Science Policy: A Federal Research Roadmap, Report on the Science of Science Policy to the Sub-Committee on Social, Behavioral and Economic Sciences, Committee on Science, National Science and Technology Council*, Washington, DC: NSTC.

NSTC (2009), *Social, Behavioral and Economic Research in the Federal Context, Sub-Committee on Social, Behavioral and Economic Sciences*, National Science and Technology Council, Washington, DC: NSTC.

Obama, Barack (2009), 'What Science Can Do', *Issues in Science and Technology*, Summer 23–30.

OECD (1992a), *OECD Proposed Guidelines for Collecting and Interpreting Technological Innovation Data – Oslo Manual*, OCDE/GD (92)26, Paris: OECD.

OECD (1992b), *Technology and the Economy: The Key Relationships*, Paris: OECD.

OECD (1997), *The World in 2020: Towards a New Global Age*, Paris: OECD.

OECD (1998), *21st Century Technologies: Promises and Perils of a Dynamic Future*, Paris: OECD.

OECD (1999a), *Economic and Cultural Transition Towards a Learning City: The Case of Jena*, Paris: OECD.

OECD (1999b), *The Future of the Global Economy: Towards a Long Boom?* Paris: OECD.

OECD (1999c), *The OECD Jobs Strategy, Implementing the OECD Jobs Strategy: Assessing Performance and Policy*, Paris: OECD.

OECD (2001a), *Measuring Productivity – OECD Manual: Measurement of Aggregate and Industry-Level Productivity Growth*, Paris: OECD.

OECD (2001b), *Science Technology Industry Review, Special Issue on New Science and Technology Indicators*, No. 27, Paris: OECD.

OECD (2001c), *Innovation and Productivity in Services*, Paris: OECD.

OECD (2001d), *The New Economy: Beyond the Hype – The OECD Growth Project*, Paris: OECD.

OECD (2002a), *Learning to Innovate: Learning Regions*, Paris: OECD.

OECD (2002b), *Frascati Manual: Proposed Standard Practice for Surveys on Research and Development*, Paris: OECD.

OECD (2003), *Measuring Knowledge Management in the Business Sector: First Steps*, Paris: OECD.

OECD (2005a), *Governance of Innovation Systems, Volume 1: Synthesis Report*, Paris: OECD.

OECD (2005b), *Governance of Innovation Systems, Volume 2: Case Studies in Innovation Policy*, Paris: OECD.

OECD (2005c), *Governance of Innovation Systems, Volume 3: Case Studies in Cross-Sectoral Policy*, Paris: OECD.

OECD (2005d), *Economic Policy Reforms 2005: Going for Growth*, Paris: OECD.

OECD (2005e), *Innovation Policy and Performance: A Cross-Country Comparison*, Paris: OECD.

OECD (2006a), *Economic Policy Reforms 2006: Going for Growth*, Paris: OECD.

OECD (2006b), *Innovation and Knowledge-Intensive Service Activities*, Paris: OECD.

OECD (2006c), *Advancing Sustainable Development*, Policy Brief, March, Paris: OECD.

OECD (2007a), *Science, Technology and Innovation Indicators in a Changing World: Responding to Policy Needs*, Paris: OECD.

OECD (2007b), *OECD Reviews of Innovation Policy: South Africa*, Paris: OECD.

OECD (2007c), *OECD Science, Technology and Industry Scoreboard 2007: Innovation and Performance in the Global Economy*, Paris: OECD.

OECD (2007d), *Economic Policy Reforms 2007: Going for Growth*, Paris: OECD.

OECD (2007e), *OECD Reviews of Innovation Policy: Chile*, Paris: OECD.

OECD (2007f), *Integrating Science and Technology into Development Policies: An International Perspective*, Paris: OECD.

OECD (2008a), 'The OECD LEED Forum on Social Innovations', http://

www.oecd.org/document/53/0,3343,en_2649_34459_39263221_1_1_1_
1,00.html, (accessed 30 September 2009).

OECD (2008b), *Productivity Measurement and Analysis*, Paris: OECD.

OECD (2008c), *OECD Reviews of Innovation Policy: Norway*, Paris: OECD.

OECD (2008d), *OECD Science, Technology and Industry Outlook 2008*, Paris: OECD.

OECD (2008e), *Main Science and Technology Indicators*, Volume 2008/2, Paris: OECD.

OECD (2008f), *The Global Competition for Talent: Mobility of the Highly Skilled*, Paris: OECD.

OECD (2008g), *Economic Policy Reforms 2008: Going for Growth*, Paris: OECD.

OECD (2008h), *Open Innovation in Global Networks*, Paris: OECD.

OECD (2008i), *OECD Reviews of Innovation Policy: China*, Paris: OECD.

OECD (2009a), *Handbook on Deriving Capital Measures of Intellectual Property Products*, Paris: OECD.

OECD (2009b), *Innovation in Firms: A Microeconomic Perspective*, Paris: OECD.

OECD (2009c), *Guide to Measuring the Information Society*, Paris: OECD.

OECD (2009d), *OECD Key Biotechnology Indicators*, Paris: OECD.

OECD (2009e), *Statistical Framework for Nanotechnology*, Paris: OECD.

OECD (2009f), *Economic Policy Reforms 2009: Going for Growth*, Paris: OECD.

OECD (2009g), *2009 Interim Report of the OECD Innovation Strategy: An Agenda for Policy Action on Innovation*, Paris: OECD.

OECD (2009h), *Growing Prosperity, Agriculture, Economic Renewal and Development, Draft Outcome Document for the Experts Meeting on 'Innovating out of Poverty'*, DCD/DAC (2009) 36, Paris: OECD.

OECD (2009i), *Green Growth: Overcoming the Crisis and Beyond*, Paris: OECD.

OECD/Eurostat (1997), *Proposed Guidelines for Collecting and Interpreting Technological Innovation Data – Oslo Manual*, Paris: OECD.

OECD/Eurostat (2005), *Oslo Manual: Guidelines for Collecting and Interpreting Innovation Data*, Paris: OECD.

OJC (2006), *Community Framework for State Aid for Research and Development and Innovation*, 2006/C323/01, Brussels: Commission of the European Communities.

OMB/OSTP (2009), *Memorandum for the Heads of Executive Departments and Agencies*, M-09-27, Washington, DC: The White House.

OST (2008), *Indicateurs de sciences et de technologies*, Paris: OST.

Parven, Sergiu-Valentin (2007), 'Community Innovation Statistics: Fourth Community Innovation Survey (CIS 4) and European Innovation Scoreboard (EIS) 2006', *Statistics in Focus, Science and Technology No. 116, 2007*, Luxembourg: European Communities.

Porter, Michael (1990), *The Competitive Advantage of Nations*, London: Macmillan.

Pro Inno Europe (2007a), *A Memorandum on Removing Barriers for a Better Use of IPR by SMEs*, A Report for the Directorate General Enterprise and Industry by an IPR Expert Group, Brussels: DG Enterprise and Industry, www.proinno-europe.eu.

Pro Inno Europe (2007b), *Guide on Dealing with Innovative Solutions in Public Procurement: 10 Elements of Good Practice*, Pro Inno Europe Paper No. 1, Commission Staff Working Document SEC (2007) 280, Brussels: DG Enterprise and Industry, www.proinno-europe.eu.

Pro Inno Europe (2009a), *European Innovation Scoreboard 2008, Comparative Analysis of Innovation Performance*, www.proinno-europe. eu/metrics, Brussels: DG Enterprise and Industry.

Pro Inno Europe (2009b), *Fostering User-Driven Innovation through Clusters*, Brussels: DG Enterprise and Industry.

Rammer, Christian, Dirk Czarnitzki and Alfred Spielkamp (2008), *Innovation Success of Non-R&D-Performers, Substituting Technology by Management in SMEs*, Discussion Paper No. 08-092, Mannheim: ZEW (Centre for European Economic Research).

République Française, Ministère de l'enseignement supérieur et de la recherche (2008), *La stratégie nationale de recherche et d'innovation chez nos concurrents*, Paris: République Française.

RICYT (2004), *Proposed Structure for an Annex to the Oslo Manual as a Guide for Innovation Surveys in Less Developed Countries*, DSTI/STP/NESTI/RD(2004)3, Buenos Aires: RICYT.

RICYT/OECD/CYTED (2001), *Standardization of Indicators of Technological Innovation in Latin American and Caribbean Countries: Bogotá Manual*, Buenos Aires: RICYT.

Rodrik, Dani (2007), *One Economics, Many Recipes: Globalization, Institutions, And Economic Growth*, Princeton, NJ: Princeton University Press.

Sainsbury, Lord (2007), *The Race to the Top: A Review of Government's Science and Innovation Policies*, Norwich: HMSO.

Sarewitz, Daniel (2003), *Does Science Policy Exist, and If So, Does it Matter? Some Observations on the US R&D Budget*, Discussion Paper for Earth Institute Science, Technology, and Global Development Seminar, 8 April.

Scerri, Mario (2006a), 'Introduction', in William Blankley, Mario Scerri, Neo Molotja and Imraan Saloojee (eds), *Measuring Innovation in OECD and Non-OECD Countries*, Cape Town: HSRC Press, pp. 1–5.

Scerri, Mario (2006b), 'The Conceptual Fluidity of National Innovation Systems: Implications for Innovation Measures', in William Blankley, Mario Scerri, Neo Molotja and Imraan Saloojee (eds), *Measuring Innovation in OECD and Non-OECD Countries*, Cape Town: HSRC Press, pp. 9–19.

Schaan, Susan and Mark Uhrbach (2009), *Measuring User Innovation in Canadian Manufacturing, 2007*, Catalogue 88F0006X, no. 3, Ottawa: Statistics Canada.

Schellings, Robert and Fred Gault (2006), *Size and Persistence of R&D Performance in Canadian Firms, 1994 to 2002*, Catalogue no. 88F0006XIE-No. 008, Ottawa: Statistics Canada.

SenterNovem (2006), *Subsidy Scheme for Innovation Vouchers*, The Hague: Ministry of Economic Affairs.

Simon, Herbert (1996), *The Sciences of the Artificial*, 3rd edn, Cambridge, MA: MIT Press.

Smith, K. (2004), 'Measuring Innovation', in J. Fagerberg, D.C. Mowery and R.R. Nelson (eds), *The Oxford Handbook of Innovation*, Oxford: Oxford University Press, pp. 148–77.

Statistics Canada (1987), 'Survey of Manufacturing Technology – June 1987', *The Daily*, 15 October, Ottawa: Statistics Canada.

Statistics Canada (1989), 'Survey of Manufacturing Technology: The Characteristics of the Plants', *Science Statistics*, **13**(10), Ottawa: Statistics Canada.

Statistics Canada (1991), *Indicators of Science and Technology 1989: Survey of Manufacturing Technology – 1989*, Catalogue 88-002, vol. 1, no. 4, Ottawa: Statistics Canada.

Statistics Canada (1999), *Science and Technology Activities and Impacts: A Framework for a Statistical Information System 1998*, Catalogue 88-522-XIB, Ottawa: Statistics Canada.

Statistics Canada (2007), *North American Industry Classification System (NAICS) 2007 – Canada*, Ottawa: Statistics Canada.

Statistics Canada (2008a), *The Canadian Research and Development Satellite Account, 1997 to 2004, Income and Expenditure Accounts Research Paper*, Catalogue no. 13-604-M no. 56, Ottawa: Statistics Canada.

Statistics Canada (2008b), 'Survey of Advanced Technology 2007', *The Daily*, 26 June, Ottawa: Statistics Canada.

Statistics Canada (2008c), 'Follow-up to the Survey of Advanced Technology 2007', *The Daily*, 27 October, Ottawa: Statistics Canada.

Statistics Canada (2009), *Industrial Research and Development: Intentions 2008*, Catalogue no. 88-202-X, Ottawa: Statistics Canada.

Sveiby, Karl-Eric (1997), *The Intangible Assets Monitor*, http://www.sveiby.com/articles/IntangAss/CompanyMonitor.html.

Uhrbach, Mark (2009), *Innovation in the Canadian Manufacturing Sector: Results from the Survey of Innovation 2005*, Catalogue 88F0006, No. 2, Ottawa: Statistics Canada.

UNESCO (2009), *Innovation for Development: Converting Knowledge to Value*, Summary Report, Paris 28–30 January 2009, Paris: UNESCO.

UNIDO (2009), *Industrial Development Report 2009, Breaking In and Moving Up: New Industrial Challenges for the Bottom Billion and the Middle-Income Countries*, E.09.ll.B.37, Vienna: UNIDO.

UN Statistical Commission (1994), *UN Fundamental Principles of Official Statistics*, http://unstats.un.org/unsd/dnss/gp/fundprinciples.aspx.

Urban, G.L. and Eric von Hippel (1988), 'Lead User Analyses for the Development of New Industrial Products', *Management Science*, **34**(5), 569–82.

US Census Bureau (2009), *Business R&D and Innovation Survey*, Washington, DC: US Census Bureau.

US Department of Commerce (1989), *Manufacturing Technology 1988*, Current Industrial Reports, Washington, DC: US Department of Commerce.

US Department of Commerce (2008), *Innovation Measurement: Tracking the State of Innovation in the American Economy*. A report to the Secretary of Commerce by the Advisory Committee on Measuring Innovation in the 21st Century Economy, Washington, DC: US Department of Commerce.

US National Science Board (2008), *Science and Engineering Indicators – 2008*, Arlington, VA: National Science Foundation.

Vespignani, A. (2009), 'Predicting the Behavior of Techno-Social Systems', *Science*, **325**, 425–8.

Vinodrai, Tara, Meric S. Gertler and Ray Lambert (2007), 'Capturing Design: Lessons from the United Kingdom and Canada', in OECD, *Science, Technology and Innovation Indicators in a Changing World: Responding to Policy Needs*, Paris: OECD, pp. 65–86.

von Hippel, Eric (1988), *The Sources of Innovation*, New York: Oxford University Press.

von Hippel, Eric (2005), *Democratizing Innovation*, Cambridge, MA: MIT Press.

von Hippel, Eric (2007), 'Democratizing Innovation: The Evolving Phenomenon of User Innovation', in OECD, *Science, Technology and*

Innovation Indicators in a Changing World: Responding to Policy Needs, Paris: OECD, pp. 125–38.

von Tunzelmann, N. (2004), 'Network Alignment in the Catching-Up Economies of Europe', in F. McGowan, S. Radosevic and N. von Tunzelmann (eds), *The Emerging Industrial Structure of the Wider Europe*, London: Routledge, pp. 23–37.

Wagner, Caroline (2008), *The New Invisible College: Science for Development*, Washington, DC: Brookings Institute Press.

Wessner, Charles W. (ed.) (2007), *Innovation Policies for the 21st Century: Report of a Symposium*, Washington, DC: National Academies Press.

Wessner, Charles W. (ed.) (2008), *An Assessment of the SBIR Program*, Washington, DC: National Academies Press.

Wolfe, David (1998), 'Social Capital and Cluster Development in Learning Regions', in J. Adam Holbrook and David A. Wolfe (eds), *Kingston: School of Policy Studies*, Kingston, ON: Queen's University.

World Bank (2008), *Global Economic Prospects: Technology Diffusion in the Developing World, 2008*, Washington, DC: World Bank.

Wright, Ronald (2004), *A Short History of Progress*, Toronto: House of Anasi Press.

Index